职业教育烹饪（餐饮）类专业“以工作过程为导向”
课程改革“纸数一体化”系列精品教材

ZHONGCAN PENGREN YINGYU

中餐烹饪英语

主　编　张艳红

参　编　（以编写章节顺序排列）
赵　静　刘欣雨　李　蕊
孙文红　王　茜　侯雪璘

華中科技大學出版社
http://www.hustp.com
中国 · 武汉

内容简介

本教材为职业教育烹饪(餐饮)类专业“以工作过程为导向”课程改革“纸数一体化”系列精品教材。

本教材共 8 个单元,内容包括厨房介绍、厨房设备与工具、调料、水果和饮品、蔬菜、畜禽肉类、水产品和中式面点。本教材用生动形象、通俗易懂的图片、对话、故事等介绍中餐烹饪英语相关知识,使教材更具可读性,同时配备英文听力材料、教学课件等丰富的数字资源。

本教材适合职业教育中餐烹饪等相关专业使用,也可作为餐饮企业员工的培训教材。

图书在版编目(CIP)数据

中餐烹饪英语/张艳红主编. —武汉:华中科技大学出版社,2020.11(2024.2 重印)
ISBN 978-7-5680-6711-9

Ⅰ. ①中…　Ⅱ. ①张…　Ⅲ. ①中式菜肴-烹饪-英语-中等专业学校-教材　Ⅳ. ①TS972.117

中国版本图书馆 CIP 数据核字(2020)第 241496 号

中餐烹饪英语

Zhongcan Pengren Yingyu　　张艳红　主编

策划编辑:汪飒婷
责任编辑:曾奇峰
封面设计:原色设计
责任校对:阮　敏
责任监印:周治超
出版发行:华中科技大学出版社(中国·武汉)　电话:(027)81321913
武汉市东湖新技术开发区华工科技园　邮编:430223
录　排:华中科技大学惠友文印中心
印　刷:武汉市洪林印务有限公司
开　本:889mm×1194mm　1/16
印　张:8
字　数:203 千字
版　次:2024 年 2 月第 1 版第 2 次印刷
定　价:39.80 元

本书若有印装质量问题,请向出版社营销中心调换
全国免费服务热线:400-6679-118　竭诚为您服务

总序

FOREWORD

职业教育作为一种类型教育，其本质特征诚如我国职业教育界学者姜大源教授提出的“跨界论”：职业教育是一种跨越职场和学场的“跨界”教育。

习近平总书记在十九大报告中指出，要“完善职业教育和培训体系，深化产教融合、校企合作”，为职业教育的改革发展提出了明确要求。按照职业教育“五个对接”的要求，即专业与产业、职业岗位对接，专业课程内容与职业标准对接，教学过程与生产过程对接，学历证书与职业资格证书对接，职业教育与终身学习对接，深化人才培养模式改革，完善专业课程体系，是职业教育发展的应然之路。

国务院印发的《国家职业教育改革实施方案》(国发〔2019〕4 号)中强调，要借鉴“双元制”等模式，校企共同研究制定人才培养方案，及时将新技术、新工艺、新规范纳入教学标准和教学内容，建设一大批校企“双元”合作开发的国家规划教材，倡导使用新型活页式、工作手册式教材并配套开发信息化资源。

北京市劲松职业高中贯彻落实国家职业教育改革发展的方针和要求，与大董餐饮投资有限公司及 20 余家星级酒店深度合作，并联合北京、山东、河北等一批兄弟院校，历时两年，共同编写完成了这套“职业教育烹饪(餐饮)类专业‘以工作过程为导向’课程改革‘纸数一体化’系列精品教材”。教材编写经历了行业企业调研、人才培养方案修订、课程体系重构、课程标准修订、课程内容丰富与完善、数字资源开发与建设几个过程。其间，以北京市劲松职业高中为首的编写团队在十余年“以工作过程为导向”的课程改革基础上，根据行业新技术、新工艺、新标准以及职业教育新形势、新要求、新特点，以“跨界”“整合”为学理支撑，产教深度融合，校企密切合作，审纲、审稿、论证、修改、完善，最终形成了本套教材。在编写过程中，编委会一直坚持科研引领，2018 年 12 月，“中餐烹饪专业‘三级融合’综合实训项目体系开发与实践”获得国家级教学成果奖二等奖，以培养综合职业能力为目标的“综合实训”项目在中餐烹饪、西餐烹饪、高星级酒店运营与管理专业的专业核心课程中均有体现。凸显“跨界”“整合”特征的《烹饪语文》《烹饪数学》《中餐烹饪英语》《烹饪体育》等系列公共基础课职业模块教材是本套教材的另一特色和亮点。大董餐饮

投资有限公司主持编写的相关教材，更是让本套教材锦上添花。

本套教材在课程开发基础上，立足于烹饪(餐饮)类复合型、创新型人才培养，以就业为导向，以学生为主体，注重“做中学”“做中教”，主要体现了以下特色。

1. 依据现代烹饪行业岗位能力要求，开发课程体系

遵循“以工作过程为导向”的课程改革理念，按照现代烹饪岗位能力要求，确定典型工作任务，并在此基础上对实际工作任务和内容进行教学化处理、加工与转化，开发出基于工作过程的理实一体化课程体系，让学生在真实的工作环境中，习得知识，掌握技能，培养综合职业能力。

2. 按照工作过程系统化的课程开发方法，设置学习单元

根据工作过程系统化的课程开发方法，以职业能力为主线，以岗位典型工作任务或案例为载体，按照由易到难、由基础到综合的逻辑顺序设置三个以上学习单元，体现了学习内容序化的系统性。

3. 对接现代烹饪行业和企业的职业标准，确定评价标准

针对现代烹饪行业的人才需求，融入现代烹饪企业岗位工作要求，对接行业和企业标准，培养学生的实际工作能力。在理实一体教学层面，夯实学生技能基础。在学习成果评价方面，融合烹饪职业技能鉴定标准，强化综合职业能力培养与评价。

4. 适应“互联网＋”时代特点，开发活页式“纸数一体化”教材

专业核心课程的教材按新型活页式、工作手册式设计，图文并茂，并配套开发了整套数字资源，如关键技能操作视频、微课、课件、试题及相关拓展知识等，学生扫二维码即可自主学习。活页式及“纸数一体化”设计符合新时期学生学习特点。

本套教材不仅适合于职业院校餐饮类专业教学使用，还适用于相关社会职业技能培训。数字资源既可用于学生自学，还可用于教师教学。

本套教材是深度产教融合、校企合作的产物，是十余年“以工作过程为导向”的课程改革成果，是新时期职教复合型、创新型人才培养的重要载体。教材凝聚了众多行业企业专家、一线高技能人才、具有丰富教学经验的教师及各学校领导的心血。教材的出版必将极大地丰富北京市劲松职业高中餐饮服务特色高水平骨干专业群及大董餐饮文化学院建设内涵，提升专业群建设品质，也必将为其他兄弟院校的专业建设及人才培养提供重要支撑，同时，本套教材也是对落实国家“三教”改革要求的积极探索，教材中的不足之处还请各位专家、同仁批评指正！我们也将在使用中不断总结、改进，期待本套教材能产生良好的育人效果。

职业教育烹饪(餐饮)类专业“以工作过程为导向”课程改革
“纸数一体化”系列精品教材编委会

职业教育烹饪（餐饮）类专业“以工作过程为导向”
课程改革“纸数一体化”系列精品教材

编委会

前言

PREFACE

《中餐烹饪英语》是中等职业学校中餐烹饪专业的英语教材，也可以作为中餐从业人员和中餐美食爱好者的自学教材。本教材遵循以就业为导向，以职业能力为本位，将行业知识和职业技能渗透在英语教学中，旨在培养学生掌握必要的专业词汇，熟悉中餐从业人员的基本技能，并通过课后练习等环节使学生可以检测自己的学习效果。本教材分成8个单元，适用于40个学时的课程教学安排。

本教材在编写时突出以下几个特色。

1. 实践性

以一位高三毕业生Jack到饭店实习中的所见所听所感为主线，以"情景教学为基石，任务为驱动，活动为内容"的方式编写，突出实践技能，强调应用，内容由浅及深，层层递进，不断提高学生的参与程度和学习兴趣，保证学习目标的实现。

2. 专业性

在参阅大量国内外中餐英语教材和深入企业调研的基础上编写而成，每个章节都由行业专家和一线中餐教师参与编写，在专业知识方面给予直接的指导。

3. 趣味性

贯穿"以学生为中心"的教学理念，设计了丰富多彩的课堂教学活动，使学生在真实的情景下进行听说读写的训练；采用大量的图片激发学生的学习兴趣，从视觉和听觉上不断吸引学生关注，激发学生学习的潜能。

本教材由北京市朝阳区教育研究中心张艳红担任主编，北京市劲松职业高中赵静、刘欣雨、王茜、孙文红、李蕊和侯雪璘参编。具体编写分工：Unit 1由赵静负责编写，Unit 2由刘欣雨负责编写，Unit 3由李蕊负责编写，Unit 4由孙文红负责编写，Unit 5由王茜负责编写，Unit 6和Unit 7由张艳红负责编写，Unit 8由侯雪璘负责编写。张艳红老师负责全书的统稿。

在本教材编写过程中，广泛吸取了国内外现有的研究成果，得到了烹饪专家和华中科技大学出版社编辑的指导和帮助，在此一并表示衷心感谢！由于学识有限，书中不妥之处敬请读者批评指正。

为了方便教学，本教材还配有链接教学课件、习题答案和英语听力材料等相关教学资源的二维码。

目录

CONTENTS

Note

Note

Unit 1

Kitchen Introduction

Learning goal（学习目标）

You will be able to：

1. know some job titles in the kitchen；
2. describe jobs in the kitchen；
3. get familiar with personal hygiene and safety in the kitchen.

扫码看课件

Lead in（导入）

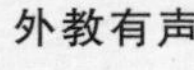

Activity 1 Self-introduction. 自我介绍。

Hello, I am Li Lei. My English name is Jack. I'm from Beijing Xinhua Vocational School. I'll be working as a trainee in the Grand Hotel for a year.

I am Wang Wei. I am a professional chef working in the Grand Hotel. I love my job! I am in charge of Jack's training now.

This is the Grand Hotel, which is a five-star hotel. The Grand Hotel serves both Chinese food and Western food.

外教有声

Activity 2 Read and translate. 读对话并翻译。

Jack：Good morning，Mr. Wang，I am Li Lei. My English name is Jack.

Chef：Nice to meet you，Jack. Welcome to the Grand Hotel.

Jack：Nice to meet you，too.

Chef：It's your first day here. Let me show you around to help you get familiar with the hotel and some rules in the kitchen.

Jack：That's so nice of you. Thank you very much.

Note

Student learning(学生学习)

Task 1 Words & expressions 词汇

Task description. 任务描述。

中餐厨房的人员分工、岗位职责、厨房的构成各不相同，任务一将聚焦于厨师的职务名称、工作内容和厨房的基本构成等相关单词和词组的学习。

Part 1 Job titles. 厨师的职务。

Activity 1 Read. 读词组。

外教有声

executive [ɪɡ'zekjʊtɪv] chef 行政总厨
executive sous-chef ['suːʃef] 行政副总厨
pastry ['peɪstrɪ] chef 中式面点厨师
assistant [ə'sɪstənt] cook 打荷厨师
cold kitchen chef 冷菜厨师
wok [wɒk] chef 炒锅厨师
chop [tʃɒp] cook 砧板厨师
water table cook 水台厨师

Activity 2 Look and write. 看图写英文。

行政总厨________

水台厨师________

炒锅厨师________

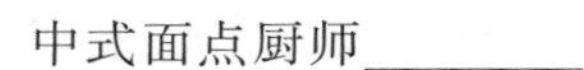

中式面点厨师________ 砧板厨师________ 冷菜厨师________

Part 2 Floor plan in the kitchen. 厨房的基本构成。

外教有声

Activity 1 Read. 读单词和词组。

kitchen [ˈkɪtʃɪn] 厨房

office [ˈɒfɪs] 办公室

locker [ˈlɒkə(r)] room 更衣室

pastry [ˈpeɪstrɪ] kitchen 面点厨房

pantry [ˈpæntrɪ] 食品储藏室

scullery [ˈskʌlərɪ] 餐具洗涤室

Activity 2 Look and write. 看图写英文。

________ ________ ________

________ ________ ________

Note

Part 3 Chef uniform. 厨师工装。

外教有声

Activity 1 Read. 读单词。

hat [hæt] 帽子

towel ['taʊəl] 手巾

jacket ['dʒækɪt] 上衣

uniform ['juːnɪfɔːm] 制服

tie [taɪ] 三角巾

apron ['eɪprən] 围裙

pants [pænts] 裤子

Activity 2 Look and Write. 看图写英文。

上衣____________

裤子____________

三角巾____________

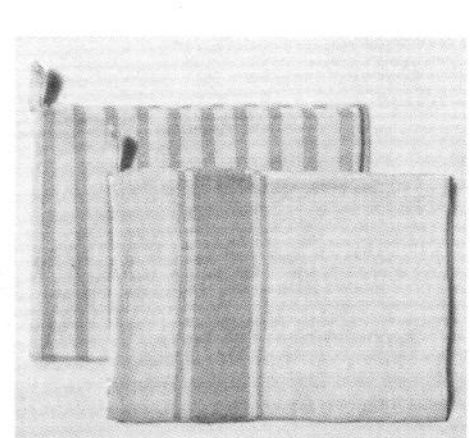

手巾____________

帽子____________

围裙____________

Activity 3 Complete. 补全单词。

1. k __ tchen 2. sc __ llery 3. ch __ f 4. a __ __ on 5. lo __ __ er

6. p __ st __ y 7. t __ w __ l 8. j __ ck __ t 9. a __ __ istant 10. ex __ c __ tive

Note

Activity 4 Making dialogues. 对话练习。

A: What does he do in the kitchen?

B: He is a /an ________.

Task 2 Personal hygiene 个人卫生

Task description. 任务描述。

卫生是厨房生产所需遵守的第一准则，包括食品卫生、员工个人卫生、环境卫生等，所有员工均应自觉遵守各项规章制度。任务二将聚焦于厨师个人卫生要求及如何保持个人卫生等英文单词和对话的学习。

外教有声

Activity 1 Read. 读单词和词组。

hygiene [ˈhaɪdʒiːn] 卫生

rinse [rɪns] 冲洗

second [ˈsekənd] 秒

wear [weə(r)] 穿、戴

wet [wet] 弄湿

soap [səʊp] 肥皂

tap [tæp] 水龙头

fingernail [ˈfɪŋgəneɪl] 手指甲

rub [rʌb] hands 搓手

dry hands 擦干手

neat [niːt] 整洁的

Note

Activity 2 Listen and complete. 听录音，补全对话。

A. rinse my hands with water	B. dry hands with a clean towel
C. rub hands for 20 seconds	D. wash your hands

Jack: Excuse me, Mr. Wang. What shall we do before cooking?

Chef: Please remember to ______________ before starting work.

Jack: Sure. I've learned the kitchen rules in my school.

Chef: That's good. Then show me how you wash hands in school.

Jack: First, wet my hands with warm water and soap, ______________. Then ______________. After turning off the tap, I will ______________.

Chef: That's correct.

Activity 3 Write. 写出下列动作的英文表述。

Activity 4 Translate. 翻译短语。

1. wet hands with soap
2. rub hands for 20 seconds
3. rinse hands with water
4. turn off the tap
5. dry hands with a clean towel

1. ______________________________
2. ______________________________
3. ______________________________
4. ______________________________
5. ______________________________

Activity 5 Match. 连线。

What else should we pay attention to while working in the kitchen?

1. Keep hair clean and neat.
2. Wear uniform and hat.
3. Keep fingernails short.
4. Not smoke, eat or drink at the work place.

A. 指甲剪短。
B. 保持头发干净、整洁。
C. 不在工作场所抽烟、吃喝。
D. 穿工装、戴厨师帽。

Task 3 Kitchen safety 厨房安全

Task description. 任务描述。

餐饮业所有人员有依法保障安全生产的责任和义务，因此所有员工必须认识到安全生产的重要性，并在工作中按要求正确操作。任务三将聚焦于厨房安全中的用刀安全、用电安全、设备安全等英文单词和对话的学习。

外教有声

Activity 1 Read. 读单词和词组。

safety [ˈseɪftɪ] 安全
besides [bɪˈsaɪdz] 除……之外(还)
rule [ruːl] 规则，制度
cause [kɔːz] 引起
attention [əˈtenʃən] 注意力
follow [ˈfɒləʊ] 遵循，跟随
avoid [əˈvɔɪd] 避免
accident [ˈæksɪdənt] 事故，意外

Note

外教有声

Activity 2 Listen and write. 听录音，写出听到的内容。

Chef：Jack，remember to pay more attention to ____________ besides your personal hygiene.

Jack：I will. I know that safety is ____________ thing while working.

Chef：Yes，that's really important.

Jack：My teacher told me I should follow the ____________ to avoid hurting myself or ____________.

Chef：That's correct. Later I will tell you some ____________.

Jack：That's so kind of you.

Activity 3 Match. 连线。

machine safety 机器安全

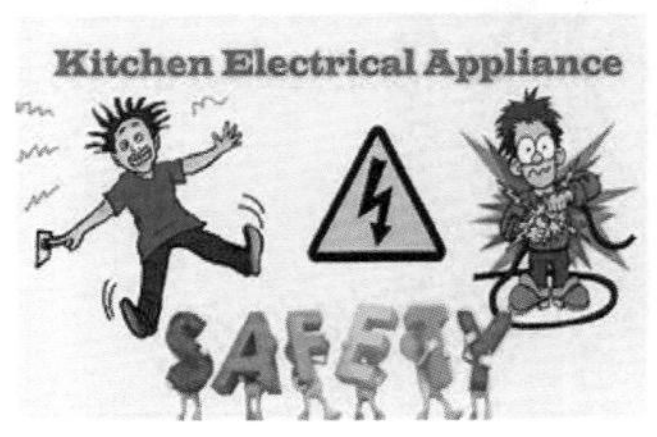

knife safety 用刀安全

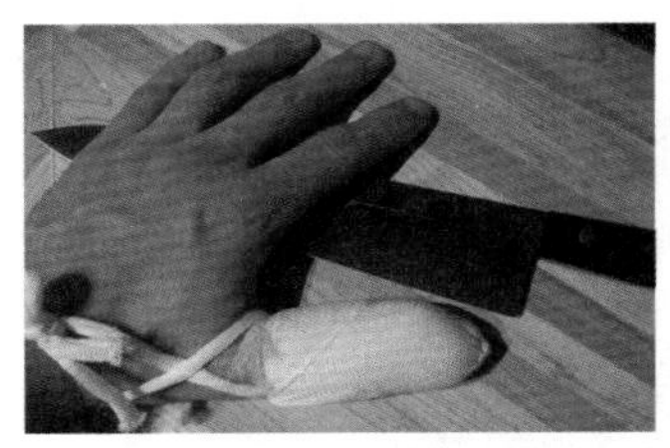

electric safety 用电安全

fire safety 用火安全

外教有声

Activity 4 Read and tick. 阅读用刀安全须知，选择正确的图片。

Keep the knife on a stable(平稳的) cutting board.

Store the knives securely(安全地) after use.

Pass a knife with the knife pointing downwards(朝下).

Do not play with a knife.

Do not stick a knife on the cutting board.

Do not touch the knife edge(刀刃) with your finger.

 □

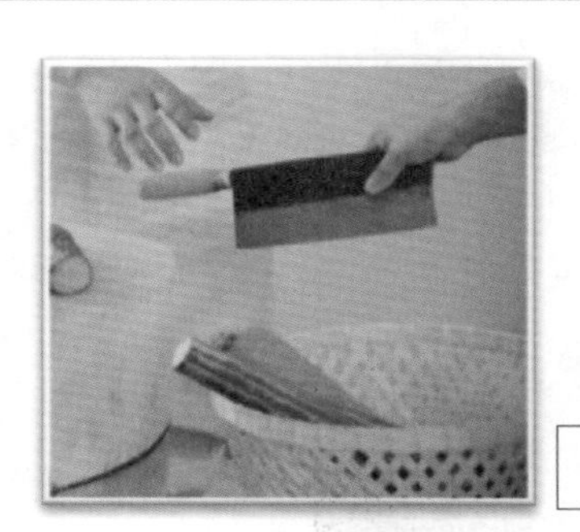 □

 □

 □

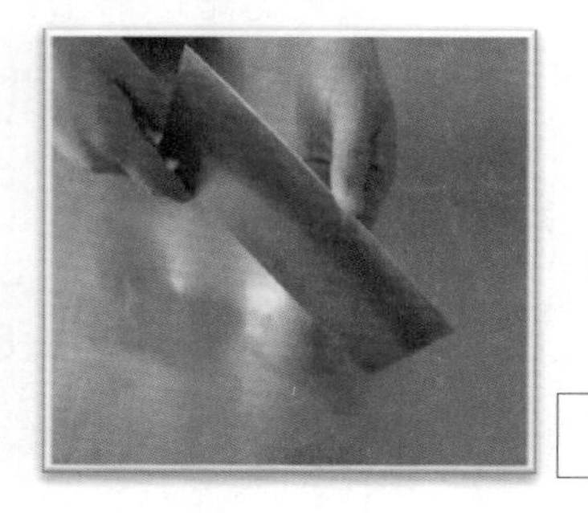 □

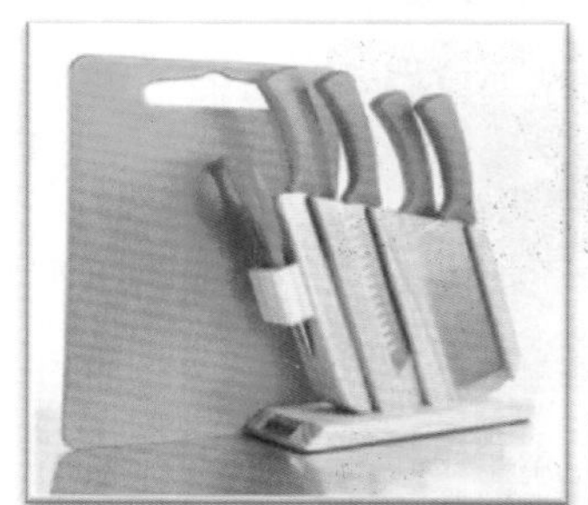 □

Culture knowledge(餐饮文化)

中餐厨师岗位简介

民以食为天，餐饮承载着历史文化的传播。中餐具有色、香、味俱全的独特魅力，有着广大的消费者和市场份额。随着其国际市场的打开，中餐厨师供不应求，而随着生活经济条件的提升，

Note

人们对美食的追求也剧增，越来越多的年轻人会选择成为中餐厨师。

中餐厨师岗位

在星级酒店，按照工作内容不同，中餐厨师岗位基本分为行政总厨、厨师长、主厨、主管、炉灶师傅、打荷、切配师傅、凉菜师傅、点心师傅等。不同的酒店对岗位的划分有一些不同，四、五星级酒店对人员分工更合理，工作岗位层级更多，包括行政总厨助理、主厨、出品总监等。

中餐厨师岗位职责

- 行政总厨：负责研发菜单，管理厨房所有事务。
- 餐厅主厨：负责烹饪、准备宴会餐盒、厨房安全卫生及人员分配。
- 砧板师傅：负责管理冰箱、验收食物及海鲜、禽畜等的宰杀。

- 凉菜师傅：冷盘、冰雕、果雕，甚至卤味都由其负责。
- 炉灶师傅：中餐的灵魂人物，负责掌灶，料理各式菜肴。
- 烧烤师傅：负责烧烤，如烤乳猪、烧腊等。
- 蒸笼师傅：炖蒸汤菜、熬制高汤都由其负责。
- 点心师傅：负责制作各式点心。
- 排菜师傅：也叫打荷，将砧板师傅切好配好的原料腌好调料、上粉上浆、用炉子烹制、协助厨师制作造型。
- 学徒：厨房职位最低，是杂役，主要工作是领取菜肉、洗菜挑菜、收拾物品和清理厨房。

Unit exercise(单元练习)

Exercise 1 Translate.(单词互译)

1. trainee ________________	2. 安全________________
3. wok chef ________________	4. 规则________________
5. kitchen ________________	6. 洗手________________
7. pastry chef ________________	8. 围裙________________
9. jacket ________________	10. 卫生________________

Exercise 2 Tick off the different words.(找不同)

1. A. wet hands	B. rub hands	C. rinse hands	D. hold water
2. A. traffic safety	B. fire safety	C. electric safety	D. food safety
3. A. chef	B. manager	C. waiter	D. cooker
4. A. clean	B. neat	C. tidy	D. dirty
5. A. glove	B. uniform	C. cook	D. pants

Note

Exercise 3 Match.（连线）

1. What hotel do you work in?
2. When should I wash my hands?
3. What do you do in the kitchen?
4. May I smoke in the kitchen?
5. What else should be done?

A. After using the toilet.
B. I work in the Great Wall Hotel.
C. Dish sink and surrounding areas should be cleaned.
D. No, you mustn't.
E. I am a pastry sous-chef.

Exercise 4 Put words in right order.（句子排序）

1. is responsible, for pastry, Mr. Huang, in the kitchen

____

2. think, to keep, the kitchen, is, clean, it, I, necessary

____

3. hands, I, will, towel, dry, with, a, clean, my

____

4. thing, is, working, the, most, important, safety, while

____

5. smoke, you, to, in the kitchen, are not allowed

____

Exercise 5 Make dialogues.（对话练习）

A: What do you do in the kitchen?
B: I am a wok chef.
A: What do you do as a wok chef?
B: I am responsible for cooking hot dishes.

pastry chef/cooking pastry
water table cook/preparing fish and seafood
cold kitchen chef/cooking cold dishes

Exercise 6 Translate.（中英文句子翻译）

1. 我在长城饭店工作。

____

2. 炒锅厨师是负责做热菜的。

____

3. 你在厨房做什么工作？我负责打荷。

____

4. Remember to pay more attention to kitchen safety besides personal hygiene.

____

5. Nobody is allowed to smoke in the kitchen.

____

Unit 1

参考答案

Note

Unit 2

Facilities & Tools

Learning goal（学习目标）

扫码看课件

You will be able to：

1. know the words and expressions of facilities and tools；
2. talk about the usages of different kinds of facilities and tools.

Lead in(导入)

Activity 1 Look and match. 看图并连线。

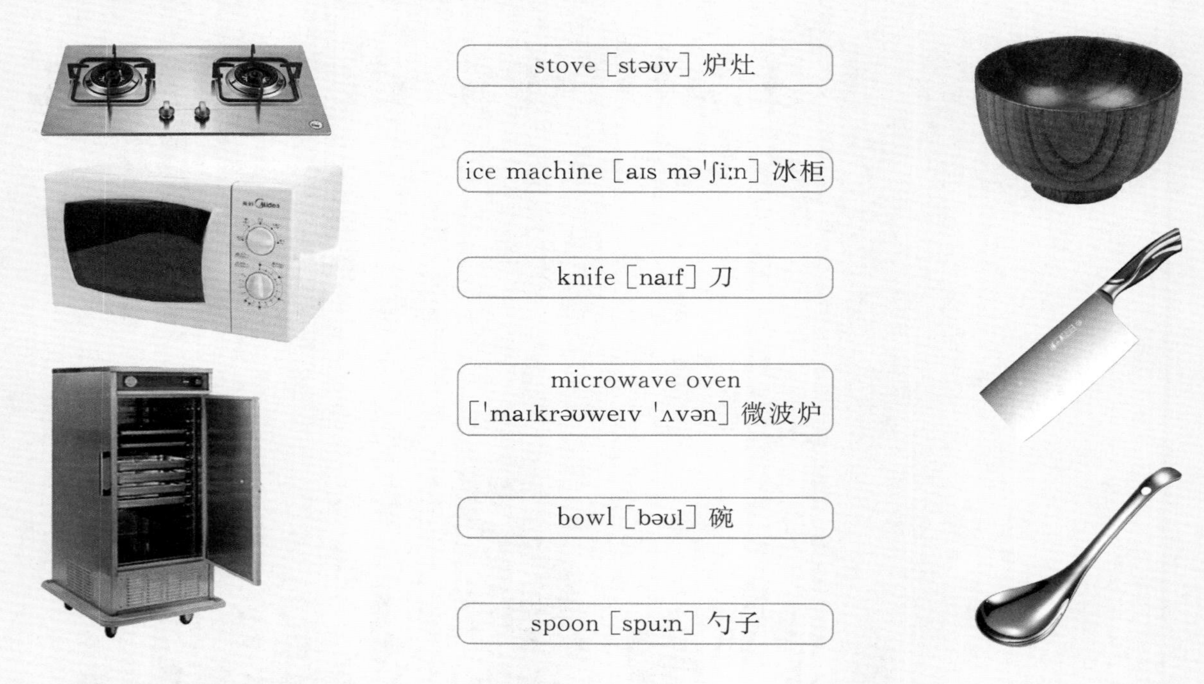

Activity 2 Decide true (T) or false (F). 判断正误。

1. ________ A microwave oven is often used to dry tableware.
2. ________ A stove is a kind of heating cookware in the kitchen.
3. ________ An ice machine is often used for making ice.
4. ________ We usually use a knife to mix ingredients.
5. ________ A spoon is often used to help people get food.

Student learning(学生学习)

Task 1 Words & expressions 词汇

Task description. 任务描述。

本章主要介绍厨房设施，其中包括大型厨房设备、常用厨房设备、厨房工具和刀具。厨房设施是帮助厨师制作精美菜肴的必备工具，中国文化讲“工欲善其事，必先利其器”。任务一主要是对大型厨房设备、常用厨房设备、厨房工具和刀具等相关单词和词组的学习。

Note

Part 1 Kitchen equipment. 大型厨房设备。

Activity 1 Read. 读单词和词组。

steamer ['stiːmə(r)] 蒸汽炉

brazier ['breɪzɪə(r)] 烤炉

fried machine [fraɪd məˈʃiːn] 炸炉

grill oven [grɪl ˈʌvən] 烤箱

steam soup stove [stiːm suːp stəʊv] 蒸汽汤炉

flathead stove with oven ['flæthed stəʊv wɪð ˈʌvən] 平头炉

steam container [stiːm kənˈteɪnə(r)] 蒸箱

heat preservation [ˌprezəˈveɪʃən] and conditioning table 保温调理台

Activity 2 Look and write. 看图写中文。

steamer ____________

brazier ____________

fried machine ____________

grill oven ____________

steam soup stove ____________

flathead stove with oven ____________

steam container ____________ heat preservation and conditioning table ____________

Kitchen facilities. 常用厨房设备。

外教有声

Activity 1 Read. 读单词和词组。

dough mixer [dəʊ 'mɪksə(r)] 和面机
mincing machine ['mɪnsɪŋ mə'ʃiːn] 绞肉机
noodle machine ['nuːdl mə'ʃiːn] 压面机
dumpling machine ['dʌmplɪŋ mə'ʃiːn] 饺子机
ice crusher [aɪs 'krʌʃə(r)] 刨冰机
electric juicer [ɪ'lektrɪk 'dʒuːsə(r)] 榨汁机
dishwasher ['dɪʃˌwɒʃə(r)] 洗碗机
blender ['blendə(r)] 搅拌机

Activity 2 Look and write. 看图写英文。

和面机________ 饺子机________ 压面机________ 搅拌机________

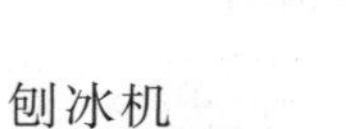

刨冰机________ 绞肉机________ 榨汁机________ 洗碗机________

Note

Part 3 Kitchen tools and utensils. 厨房工具。

外教有声

Activity 1 Read. 读单词和词组。

ladle ['leɪdl] 炒勺

rolling-pin ['rəʊlɪŋ pɪn] 擀面杖

spoon ladle [spuːn 'leɪdl] 汤勺

shovel ['ʃʌvəl] 铲子

wok [wɒk] 中式炒锅

spider ['spaɪdə(r)] 漏勺

meat mallet [miːt 'mælɪt] 肉锤

seasoning pot ['siːzənɪŋ pɒt] 调味罐

Activity 2 Write and translate. 单词互译。

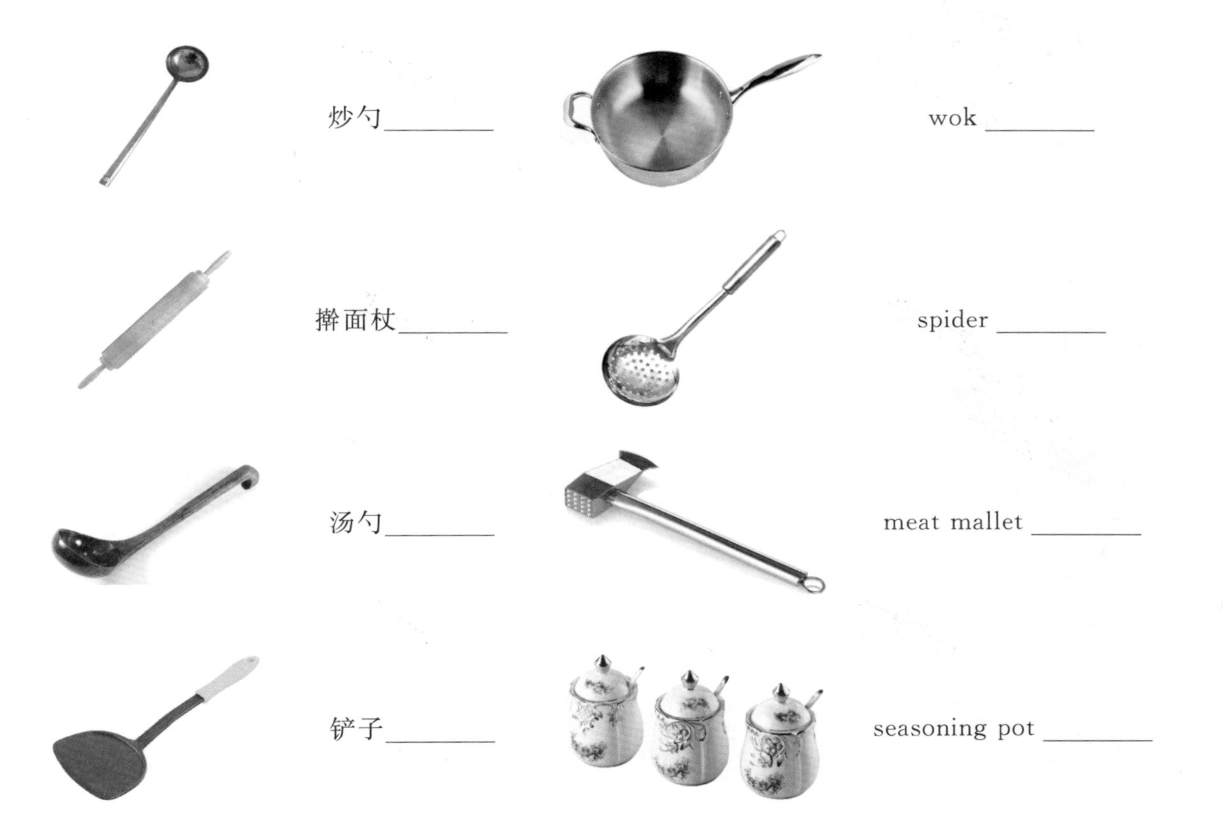

炒勺________ wok ________

擀面杖________ spider ________

汤勺________ meat mallet ________

铲子________ seasoning pot ________

Note

Part 4 Kitchen knives. 厨房刀具。

外教有声

Activity 1 Read. 读单词和词组。

chopping knife [ˈtʃɒpɪŋ naɪf] 文武刀
cleaver [ˈkliːvə(r)] 砍刀
peeler [ˈpiːlə(r)] 刮皮刀
carving knife [ˈkɑːvɪŋ naɪf] 雕刻刀
slicing knife [ˈslaɪsɪŋ naɪf] 片刀
slices of duck sword [sɔːd] 片鸭刀

Activity 2 Write and translate. 单词互译。

carving knife ________

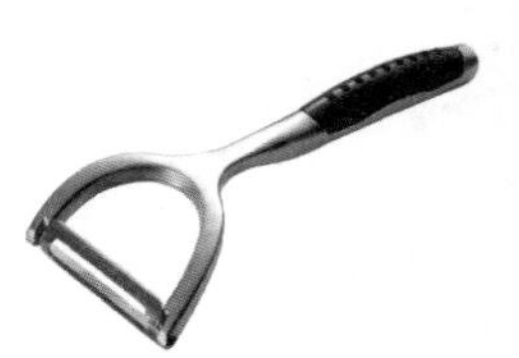
刮皮刀________

cleaver ________

片刀________

chopping knife ________

片鸭刀________

Activity 3 Complete. 补全单词。

1. sh __ vel 2. sp __ der 3. sl __ cing knife 4. sp __ __ n 5. s __ __ soning p __ t
6. __ leaver 7. c __ __ ving knife 8. w __ k 9. mall __ t 10. ch __ pping knife

Note

Activity 4 Making dialogues. 对话练习。

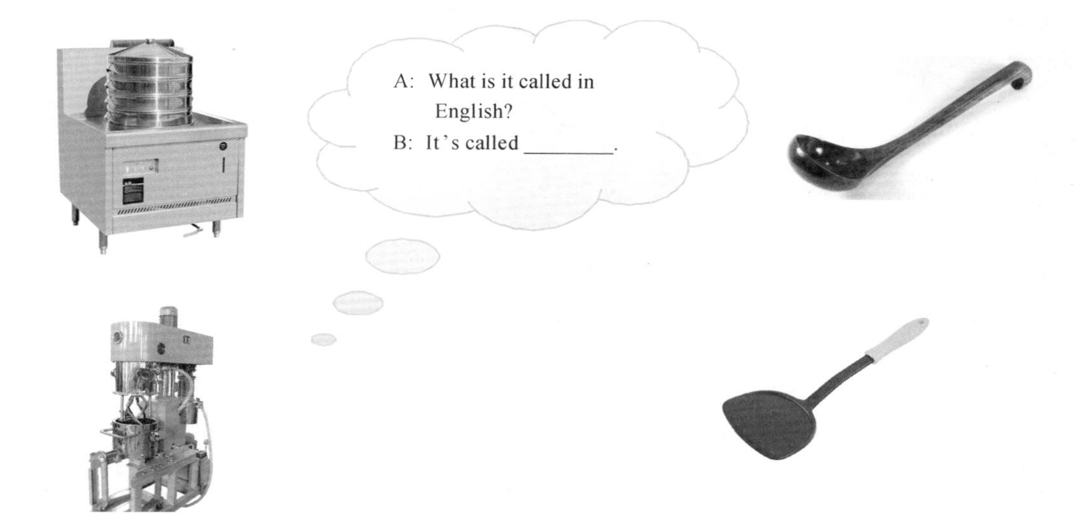

Task 2 Application of kitchen tools 厨房工具的应用

Task description. 任务描述。

任务二将聚焦于各类厨房设备、工具等的应用以及它们的操作技法的英文表达。

Part 1 Application of kitchen equipment. 厨房设备的应用。

Activity 1 Read. 读单词和词组。

外教有声

machine [mə'ʃiːn] 机器　tidy ['taɪdɪ] 干净的　serving ['sɜːvɪŋ] 上菜

dough [dəʊ] 面团　press [pres] 按下　switch button [swɪtʃ 'bʌtən] 电源按钮

keep dishes warm 让菜保温

Activity 2 Listen and complete. 听录音，补全对话。

外教有声

A. keeping dishes warm before serving

B. heat preservation and conditioning table

C. put the dough on top of the machine

D. press the red switch button

E. we usually use it to make different noodles

Chef：Today I'll show you around our kitchen.

Jack：Really? That's great.

Chef：OK，let's look at the machine. It's called steamer.

Jack：Wow! It's so tidy. How about that?

Chef：It's ________________.

Jack：What is it used for?

Chef：It is used for ________________.

Jack：Is it a noodle machine?

Chef：Yes，________________.

Jack：I see. How should I use it?

Chef：Let me show you. First，________________. Then ________________. You'll see noodles coming out of the noodle machine.

Jack：That's great! Oh，I learn a lot today. Thank you.

Activity 3 Write. 写出对话中所提及设备的英文名称。

1. ________________ 2. ________________ 3. ________________

Activity 4 Look and choose. 看图并选择。

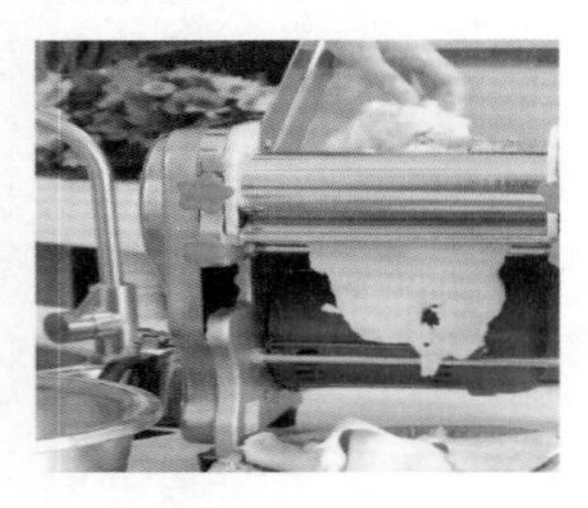

____

A. You'll see noodles coming out of it.

____

B. Put the dough on the top of the machine.

C. Press the red switch button.

____

Part 2 Application of tools & utensils. 常用工具的应用。

外教有声

Activity 1 Read. 读单词和词组。

cookware ['kʊkweə(r)] 炊具
commonly ['kɒmənlɪ] 通常
a large surface area 巨大的表面积
be suitable ['suːtəbəl] for 对……合适
increase [ɪn'kriːs] 增多
keep... fresh taste 保持新鲜的味道
save oil 省油

外教有声

Activity 2 Listen and write. 听录音，写出听到的内容。

Jack: Chinese wok is one of the most important ________ in the kitchen. Could you tell me why ________ are used so much in Chinese cooking?

Chef: Well, for some reasons, for example, the wok has ________ to increase the speed.

Jack: Are there any other reasons?

Chef: It can keep almost vegetables' ________.

Jack: Is it so important?

Chef: Yes. At the same time, the wok's ________ help to keep the food in the rounded bottom, so it can ________.

Jack: I see. Excuse me, sir. Can we use a wok for boiling?

Chef: A wok is most ________ used for stir frying. We steam dumplings or other food with steamers. Soup pots are suitable for stewing and making soup.

Jack: I've really learned a lot. Thank you so much.

Activity 3 Look and describe. 看图并参考 Activity 2 用英文描述。

炒锅外形描述________

Note

炒锅保鲜描述________________

炒锅省油描述________________

Activity 4 Read and judge. 阅读 **Activity 2**，判断下列说法是否正确。

1. We can use a wok for making soup. (　　)
2. We can use a wok for steaming dumplings. (　　)
3. A wok is most commonly used for stir frying. (　　)

Part 3 Application of knives. 刀的应用。

外教有声

Activity 1 Read and describe. 朗读单词，用英文描述下列图片。

cut [kʌt] 切，切菜	peel [piːl] 削皮	dice [daɪs] 切块
flatten ['flætən] 拍平	chop [tʃɒp] 剁，砍	slice [slaɪs] 切片
shred [ʃred] 切丝	mince [mɪns] 切末，切碎	

________________　________________　________________　________________

________________　________________　________________　________________

Note

外教有声

Activity 2 Read and write. 读文章并写出三种磨刀方式。

A：My knife is dull. It doesn't work well.

B：Use a steel.

A：Where can I find it?

B：You can use mine. Here you are.

A：Thank you. What else can I use to sharpen my knife?

B：You may use a whetstone, or you can also use an electric knife sharpener. It is safer.

A：Oh, I see. Thank you for telling me.

1. ____________

2. ____________

3. ____________

Part 4 Application of other tools. 其他工具的应用。

外教有声

Activity 1 Read. 读单词和词组。

steamed bun [bʌn] 馒头 pastry ['peɪstrɪ] 糕点 braising [breɪzɪŋ] pan 炖锅

low heated dish [ləʊ 'hiːtɪd dɪʃ] 小火慢炖的菜肴 marmite ['mɑːmaɪt] 砂锅

wax gourd [wæks gʊəd] 冬瓜 melon ['melən] baller 挖球器

special ['speʃəl] 特殊的 vessel ['vesəl] 容器 holding ['həʊldɪŋ] soup 盛汤

外教有声

Activity 2 Read the dialogue. 读对话。

Jack：What kind of pot is the best for steamed bun?

Chef：The steamer, which is used for steaming pastries.

Jack：How about marmite and braising pan?

Chef：They are suitable for the low heated dishes.

Jack：How do you make these round wax gourd balls?

Chef：Melon baller helps us to make it.

Jack：It is wonderful! So the large bowl of wax gourd soup looks beautiful. What is the special vessel for?

Chef: It is a jar. It is used for holding soup.

Activity 3 Match. 连线。

stir-fry the vegetables

press the dough

fish out dumplings

season with salt and sugar

soften the meat

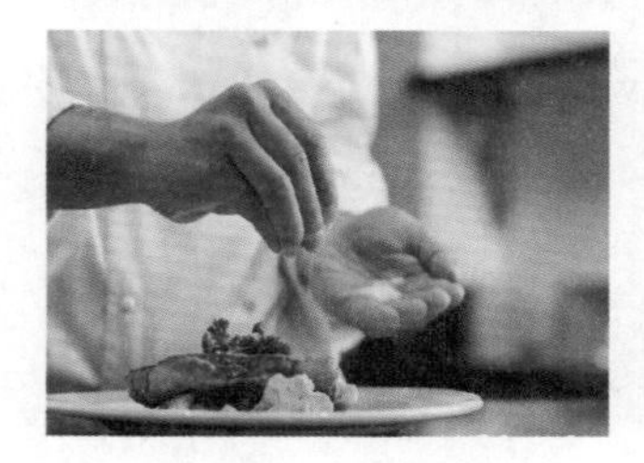

Activity 4 Write and translate. 写出下面工具的英文名称并翻译其功能。

1. ____________

它适合烹饪小火加热的菜肴。

____

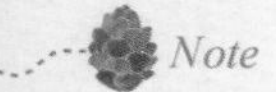

2. ______________

蒸笼是蒸小笼包最好的工具。

____

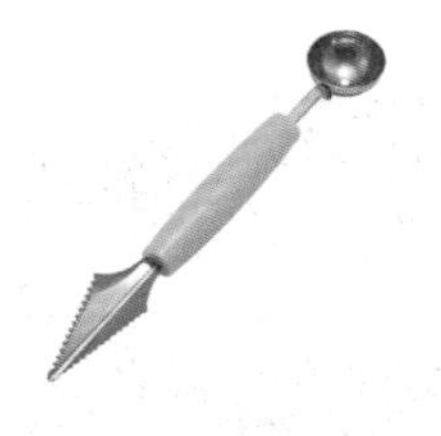

3. ______________

Melon baller helps us to make balls.

____

4. ______________

It is used for holding soup.

____

Culture knowledge(餐饮文化)

中餐餐桌礼仪基本常识

中国的餐饮文化历史悠久,中餐的餐桌礼仪更是源远流长,礼仪体现了一个人内在的修养与素质,也是人际交往中的一种适用艺术,示人以尊重、友好。下面就向大家介绍中餐餐桌礼仪基本常识,使大家在交往中做到得体而不失礼仪。

❶ 筷子文化

筷子虽然用起来简单、方便,但也有很多规矩。比如:不能举着筷子和别人说话,说话时要将筷子放到筷架上,或将筷子并齐放在饭碗旁边。不能用筷子去推饭碗、菜碟,不要用筷子去叉馒头或别的食品。

❷ 餐巾文化

正式宴会前,每位用餐者面前会有一条湿毛巾,它是用来擦手的,不能当作他用。之后,再为用餐者准备一条餐巾,应将它铺放在并拢的大腿上,不可围在别处。餐巾可以用于轻抹嘴部和手,不能擦餐具或擦汗。

❸ 中餐上菜的顺序

标准的中餐,无论何种风味,其上菜顺序大体相同。通常首先是冷盘,接着是热炒,随后是主菜,然后上点心、汤,最后是水果拼盘。当冷盘吃到1/3时,开始上第一道菜,一般每桌安排10个热菜。服务员上菜按先主宾后主人、先女士后男士的顺序。

以主宾位置沿顺时针方向取菜，不可迫不及待。

④ 用餐的礼仪

上菜后，不要先拿筷子，应等主人邀请，主宾拿筷子时再拿筷子。取菜时要相互礼让，不要争抢，取菜时要适量。不要挑菜或翻菜，看准后要立即取走，不能再放回去。

⑤ 席位文化

①右高左低原则：两人一同并排就座用餐，通常以右为上座，以左为下座。这是因为中餐上菜时多以顺时针方向为上菜方向，居右坐的要比居左坐的优先受到照顾。

②中座为尊原则：三人一同就座用餐，坐在中间的人在位次上高于两侧的人。

③面门为上原则：用餐的时候，按照礼仪惯例，面对正门者是上座，背对正门者是下座。

Unit exercise(单元练习)

Exercise 1 Translate.（单词互译）

1. steamer ________________　　2. 炒勺________________

3. brazier ________________　　4. 汤勺________________

5. fried machine ________________　　6. 中式炒锅________________

7. grill oven ________________　　8. 漏勺________________

9. peeler ________________　　10. 砍刀________________

Exercise 2 Tick off the different words.（找不同）

1. A. chop　B. slice　C. mince　D. cutting

2. A. steamer　B. brazier　C. rolling-pin　D. fried machine

3. A. season　B. press　C. dough　D. soften

4. A. cleaver　B. carving knife　C. chopping knife　D. small knife

5. A. sharpen knives　B. cutting board　C. stick a knife　D. store the knives

Exercise 3 Match.（连线）

1. mince meat with　　A. a mallet

2. sharpen a knife with　　B. a melon baller

3. beat the meat with　　C. a mincing machine

4. hold soup with　　D. a spoon ladle

5. make watermelon balls with　　E. a whetstone

Exercise 4 Choose.（选择正确的动词）

A. ________ the pan off the fire. Boil

B. ________ the meat in the hot pot. Take

C. ________ the plate with the anti-hot folder. Grate

D. ________ the surface of the oil with a soup ladle. Lift

E. ________ the potato with the grater. Clean

Exercise 5 Put words in right order.（句子排序）

1. the knife, on, keep, a stable cutting board

____

2. dull, the knife, is, with, you, the steel, sharpen, it

____

3. ice crusher, we, break, usually, in summer, ice, with

____

4. carving knives, cooks, carve, use, for, food, decorating, dishes, to

____

5. equipment, steam soup stove, in the kitchen, the most important, is, one of

____

Exercise 6 Translate.（将用食材制作食品的工具翻译成英文）

1. 我们用榨汁机榨橙汁。

____

2. 厨师用片鸭刀片鸭肉。

____

3. 绞肉机使绞肉变得更容易。

____

Exercise 7 Translate.（中英文句子翻译）

1. 我能用炒锅做汤吗？

____

2. 请用这把砍刀将猪肉剁开。

____

3. 杰克正在用和面机和面。

____

4. What else can I use to sharpen my knife?

____

5. Clean the surface of the oil with a soup ladle.

____

Unit 2
参考答案

Unit 3

Seasonings

You will be able to：

1. read and write the words of seasonings；
2. talk about the ways of using spices and condiments.

扫码看课件

Lead in(导入)

Activity 1 Look and match. 看图并连线。

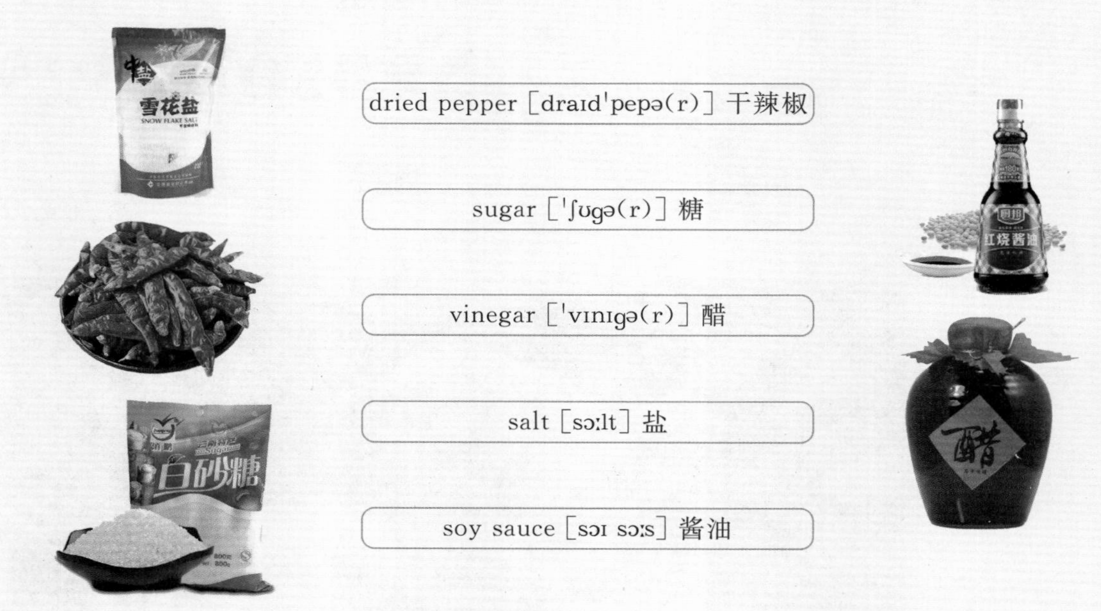

Activity 2 Decide true (T) or false (F). 判断正误。

1. ________ Dried pepper is widely used in Hunan.
2. ________ We don't use salt when we use soy sauce.
3. ________ We can use soy sauce instead of vinegar.
4. ________ Sugar must be used in sweet and sour food.
5. ________ Many people in different areas like to add sugar to cold dishes.

Student learning(学生学习)

Task 1 Words & expressions 词汇

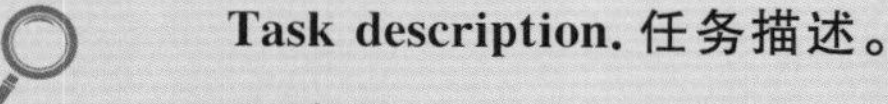

Task description. 任务描述。

调料是人们用来制作食品等的辅助用品。调料分为干货类、腌菜类、鲜类、酿造类、水产类和其他。任务一将聚焦于这些种类调料的单词和词组的学习。

Note

Part 1 Dry seasonings. 干货类调料。

外教有声

 **Activity 1** Read. 读单词和词组。

wild [waɪld] pepper 花椒	mustard ['mʌstəd] 芥末
bay [beɪ] leaf 香叶	pepper powder ['paʊdə(r)] 胡椒粉
fennel ['fenəl] 茴香	chili powder 辣椒粉
aniseed ['ænɪsiːd] 八角	cinnamon ['sɪnəmən] 桂皮

Activity 2 Look and write. 看图写中文。

bay leaf

____

pepper powder

____

chili powder

____

wild pepper

____

cinnamon

____

fennel

____

aniseed

____

mustard

____

Part 2 Preserved & fresh seasonings. 腌菜类和鲜类调料。

外教有声

 **Activity 1** Read. 读单词和词组。

preserved [prɪ'zɜːvd] vegetable 梅干菜	garlic ['gɑːlɪk] 蒜
pickled ['pɪkld] pepper 泡辣椒	scallion ['skæljən] 葱
chili ['tʃɪlɪ] 辣椒	ginger ['dʒɪndʒə(r)] 姜
horseradish ['hɔːsˌrædɪʃ] 辣根	

Note

Activity 2 Look and write. 看图写英文。

葱 ____________　　蒜 ____________　　梅干菜 ____________　　辣根 ____________

姜 ____________　　辣椒 ____________　　泡辣椒 ____________

Part 3　Paste & seafood seasonings. 酿造类和水产类调料。

外教有声

Activity 1 Read. 读词组。

shrimp paste [ʃrɪmp peɪst] 虾酱

sweet sauce [sɔːs] 甜面酱

soya bean [ˈsɔɪə biːn] paste 豆瓣酱

oyster [ˈɔɪstə(r)] oil 蚝油

bean paste 黄酱

dried shrimp 虾皮

fish sauce 鱼露

Activity 2 Write and translate. 单词互译。

虾皮 ____________

fish sauce ____________

蚝油 ____________

bean paste ____________

Note

虾酱________

soya bean paste ________

甜面酱________

Part 4 Other seasonings. 其他类。

外教有声

Activity 1 Read. 读单词和词组。

sesame ['sesəmɪ] oil 芝麻油	MSG 味精
rice wine 黄酒	peanut ['piːnʌt] oil 花生油
chili oil 辣椒油	ketchup ['ketʃəp] 番茄酱
cornstarch ['kɔːnstɑːtʃ] 淀粉	five-spice powder 五香粉

外教有声

Activity 2 Listen and write. 听录音，写单词。

____

____

____

____

____

____

____

____

Activity 3 Making dialogues. 对话练习。

A: What kind of seasonings
will you often use
during cooking?
B: We often use ________ .

Task 2 Usage of seasonings 调料的应用

Task description. 任务描述。

将调料少量加入食物中可以改善食物的味道。从调料所呈现的味道上可将其分为酸、甜、辣、咸、鲜等，调料在烹饪过程中的使用时间不同会带来不同的效果，如淀粉的使用时间。任务二将聚焦于各种调料的应用及调料的计量的英文表达。

Part 1 Usage of dry seasonings. 干货类调料的应用。

外教有声

Activity 1 Read. 读单词和词组。

pickled [ˈpɪkəld] food 卤菜
sprinkle [ˈsprɪŋkl] 淋，洒
widely [ˈwaɪdlɪ] 广泛地
especially [ɪˈspeʃəlɪ] 特别地

Note

Activity 2 Listen and complete. 听录音，补全对话。

A. add some wild pepper	B. sprinkle some sesame oil
C. is widely used in hot dishes	D. in pickled and stewed food

Jack: Excuse me. Could you tell me how to use pepper powder, chef?

Chef: Pepper powder ________, and add it just before the dish is ready.

Jack: I know. How about aniseed, fennel, cinnamon and bay leaf?

Chef: They are often used ________.

Jack: Is it important to ________ and chili to make spicy food?

Chef: Yes. Especially Sichuan and Chongqing food. It's more delicious to ________.

Part 2 Usages of cornstarch. 淀粉的应用。

Activity 1 Read. 读词组。

cover ['kʌvə(r)]... with 裹

make... crispy ['krɪspɪ] 使……酥脆

make... thicken ['θɪkən] (使)变浓稠

mix with 与……混合

Activity 2 Listen and order. 听录音并排序。

Chef: ____
Jack: ____
Chef: ____
Jack: ____
Chef: ____
Jack: ____
Chef: ____
Jack: ____
Chef: ____
Jack: ____

A. But cornstarch has different usages.

B. OK. Why do you put it in the chicken?

C. Cover the meat with starch, and it will make the chicken crispy when you fry it.

D. Oh. I know.

E. Why do they put it in dishes?

F. Yes. Cornstarch is widely used in hot dishes.

G. They put it just before the dish is done. It will make the sauce thicken.

H. Hi, Jack. Could you pass me the cornstarch?

I. I only saw somebody use the cornstarch mixed with water.

J. I got it. Thank you!

Activity 3 Look and match. 看图并连线。

A. Cover the meat with cornstarch, and get ready to fry.

B. When you preserved the meat, put some cornstarch, it will make the meat soft.

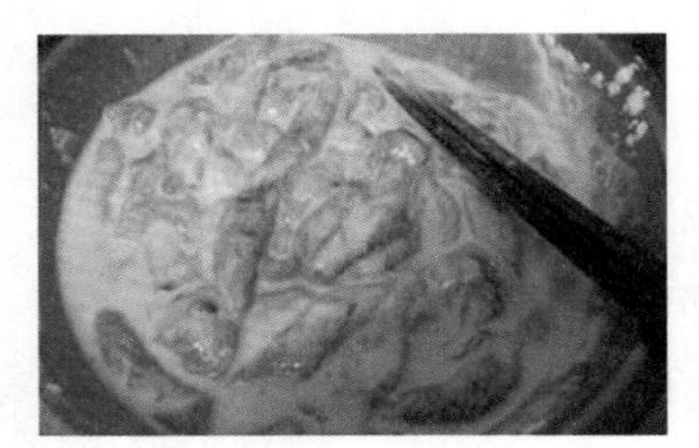

C. Put some water in cornstarch, and stir it frequently.

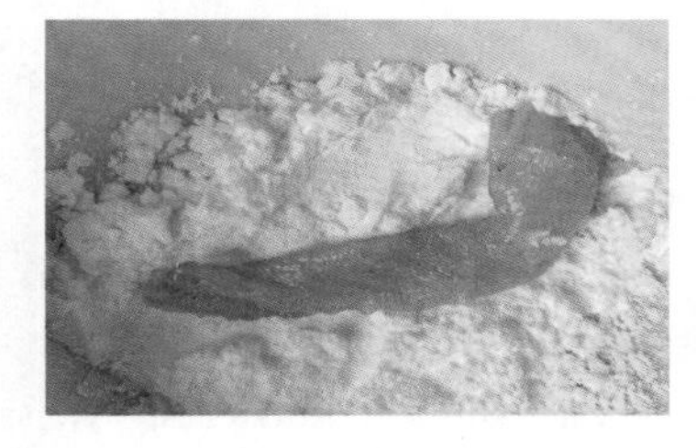

D. Put water cornstarch to thicken the sauce before dish up.

Part 3 Usage of oil. 食用油的应用。

外教有声

Activity 1 Read. 读单词和词组。

corn [kɔːn] oil 玉米油

olive [ˈɒlɪv] oil 橄榄油

deep-fry [ˈdiːpˈfraɪ] 深炸

complicated [ˈkɒmplɪkeɪtɪd] 复杂的

Activity 2 Listen and write. 听录音，写出听到的内容。

Chef: Jack! Pass me the oil!

Jack: Yes. There are many kinds of oil. ____________, olive oil, peanut oil or corn oil?

Chef: Corn oil. I ______________ the chicken.

Jack: Why don't you choose peanut oil?

Chef: The food will be greasy if you use peanut oil to fry.

Jack: ________________ olive oil?

Chef: We always use it when we __________________.

Jack: There is sesame oil, too.

Chef: Sesame oil is usually used in cold dishes, and it will make the dish delicious.

Jack: Wow. That's too complicated.

Activity 3 Write. 看图并写出油的名称和主要用途。

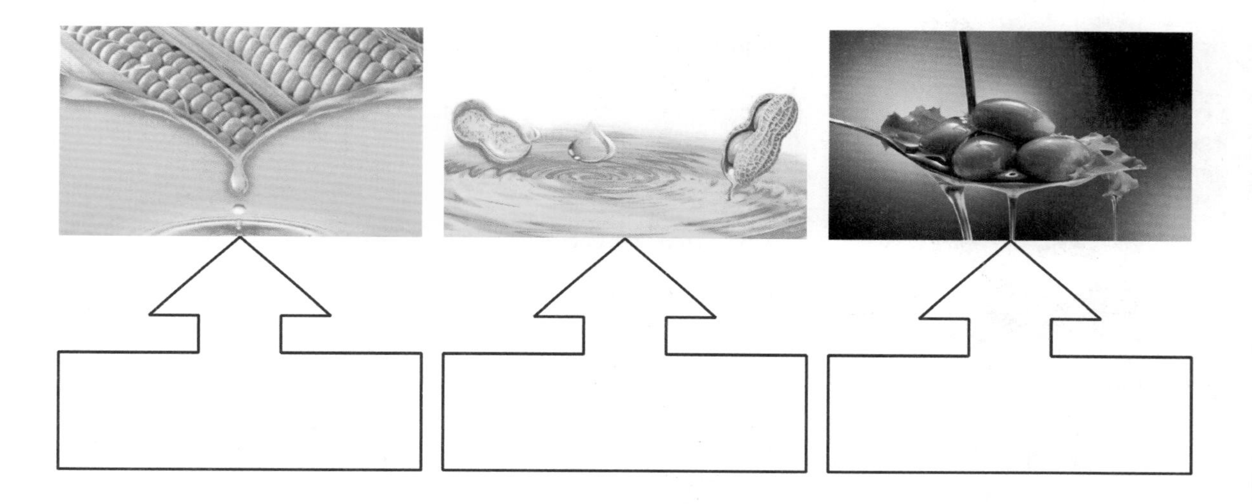

Part 4 Amount of seasonings. 调料的计量。

Activity 1 Look and match. 量词的缩写形式和中文连线。

tea spoon	g	汤匙
gram	L	毫升
table spoon	mL	茶匙
liter	tsp	克
milliliter	tbsp	升

Activity 2 Look and translate. 看图并翻译。

1. People need to drink at least 2 L water every day.

____

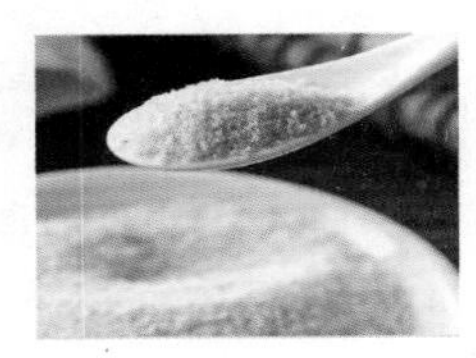

2. Help me to get a tea spoon of salt, please.

____

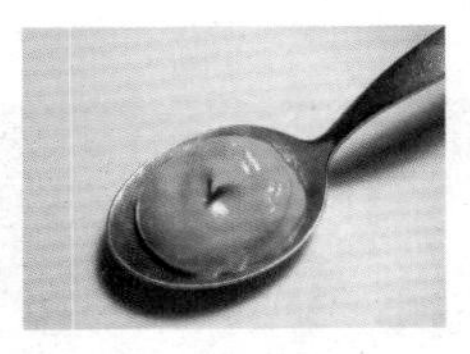

3. You need to add a table spoon of mustard.

____

4. Five dried peppers weigh about 5 g.

____

Task 3 Making dishes 菜肴制作

Task description. 任务描述。

不同调料组合在一起能形成不同的口味，从而使每一类食品都有自己的特色。重庆火锅在制作配料上极具特色，任务三聚焦于重庆火锅主料、调料和制作过程英语描述的学习。

外教有声

Activity 1 Read the recipe. 读菜单。

Chongqing hotpot(重庆火锅)

Ingredients(原料)

red dried pepper　ginger　wild pepper　salt

garlic　Pixian soy bean paste　salad oil

chili sauce

Note

Methods (制作方法)

1. Fry the red dried peppers in a pot and mince them all.
2. Put salad oil and the minced chili into the pot.
3. Add the left seasonings in the pot. Stir-fry the mixture for over two hours.

Characteristics (特点)

It is the most famous dish in Chongqing, with spicy and hot. People enjoy the delicious base and dipping sauce.

Activity 2 Decide true (T) or false (F). 判断正误。

1. ________ We need not cut pepper before fry it.
2. ________ Pixian soy chili paste is widely used in Sichuan dish.
3. ________ We don't use salt when we cook hotpot.
4. ________ We stir-fry the mixture for just one hour.
5. ________ We put ginger, garlic and scallion in hotpot.

Activity 3 Translate. 中英文互译。

Ingredients(原料)

调和油____________

chili sauce ________

干辣椒__________

ginger ____________

花椒____________

garlic ____________

郫县豆瓣酱________

salt ____________

Methods(制作方法)

1. Heat the pot first, and then fry the red dried peppers. Mince the peppers to small pieces.

____

____

Note

2. Put the fried pepper into the pot, and then add enough salad oil.

____

____

3. Add Pixian soy bean paste, wild peppers and other ingredients into the pot. Stir-fry the mixture for over two hours.

____

____

Culture knowledge(餐饮文化)

酱油的由来

中式烹饪中除了百味之首的盐外，重要性排在第二位的就要数酱油了。

酱油是中国人发明的，据说，早在西周的时候，古人就知道把新鲜的肉弄碎发酵，做成肉酱，元明时期，出现了用大豆制成的酱汁。至清代，酱油的口味已经和我们今天吃到嘴里的酱油的味道相差无几了。清代《醒园录》中记载的酱油做法十分繁复，没有半年的工夫，是酿造不出来的。

大豆蛋白质在蛋白酶和肽酶作用下，经水解生成小分子的肽，产生一定的鲜味；在蛋白酶的作用下，继续分解为 18 种氨基酸，其中谷氨酸与天冬氨酸能够提供更为浓郁的鲜味，而酪氨酸氧化成棕色黑色素，令酱油变成黑色。

酱油与盐最大的区别是酱油除了咸还有鲜，是通过自然发酵，历经沉淀才能得到的鲜美之味。

成年人每天油盐的健康摄入量

❶ 盐的摄入量

控盐是高血压等慢性病预防的综合干预的重要措施之一，体现了一种健康的生活方式。《中国居民膳食指南(2016)》推荐成年人每日食盐摄入量不超过 6 g，约相当于一啤酒瓶盖的量。同时还要注意隐性钠的问题，少吃高盐食品，酱油、咸菜、酱豆腐、味精、鸡精都是常见的隐藏“盐”，面条、面包、饼干等在加工过程中，都添加了食盐。

Note

❷ 油的摄入量

食用油是人体70%的脂肪酸来源和人体重要的基本营养素之一，但食用油含饱和脂肪酸过多，会引起胆固醇增高，高血压、冠心病、糖尿病、肥胖症等疾病也就会相继“造访”。《中国居民膳食指南(2016)》指出成年人每天的食用油摄入量不宜超过25 g，不但包括自家烧菜所用的食用油，还包括食用猪肉等食品所摄入的油脂。

Unit exercise(单元练习)

Exercise 1 Translate.(单词互译)

1. olive oil ________________ 2. 香叶________________
3. rapeseed oil ________________ 4. 辣椒粉________________
5. aniseed ________________ 6. 豆瓣酱________________
7. soy sauce ________________ 8. 花生油________________
9. cinnamon ________________ 10. 淀粉________________

Exercise 2 Tick off the different words.(找不同)

1. A. vinegar B. oil C. rice D. soy sauce
2. A. ketchup B. mustard C. egg D. bean paste
3. A. salt B. sugar C. vinegar D. corn
4. A. sweet sauce B. cinnamon C. beef D. chili oil
5. A. coffee B. mustard C. bay leaf D. cornstarch

Exercise 3 Match.(连线)

A. Could you pass me the chicken powder? 1. Mix the salt, vinegar, MSG, cornstarch, sugar, minced ginger and so on.

B. Do you have any salad oil? 2. Yes, of course.

C. How much do you want? 3. A table spoon, please.

D. How can I use the sesame oil? 4. Sprinkle it on the duck.

E. Can you tell me how to make the sauce? 5. Sure. Here you are.

Exercise 4 Put words in right order.(句子排序)

1. like, to, dishes, add, sesame oil, to, some, the, people, cold

Note

2. is, often, mustard, with, sea food, served

3. to, like, dumpling, vinegar, they, go with

4. salt, you, some, soup, put, the, have, in

5. do not, eat, hotpot, want, chili, I, when, I, sauce

Exercise 5 Write.（区分下列调料并写出名称）

Exercise 6 Translate.（中英文句子翻译）

1. 盘中滴几滴香油。

2. 汤里加点胡椒粉，这样会更加鲜美。

3. 这道凉菜用葱、姜、蒜、酱油、味精、醋及白砂糖调和而成。

4. Pass me a glass of white wine, please.

5. Boil the red dry peppers in a pot and mince them all.

Unit 3

参考答案

Unit 4

Fruits & Drinks

Learning goal（学习目标）

扫码看课件

You will be able to：

1. know the words and expressions of fruits and drinks；
2. describe the ways of preparing fruits and drinks；
3. talk about the steps of making fruit dishes.

Lead in(导入)

Activity 1 Look and match. 看图并连线。

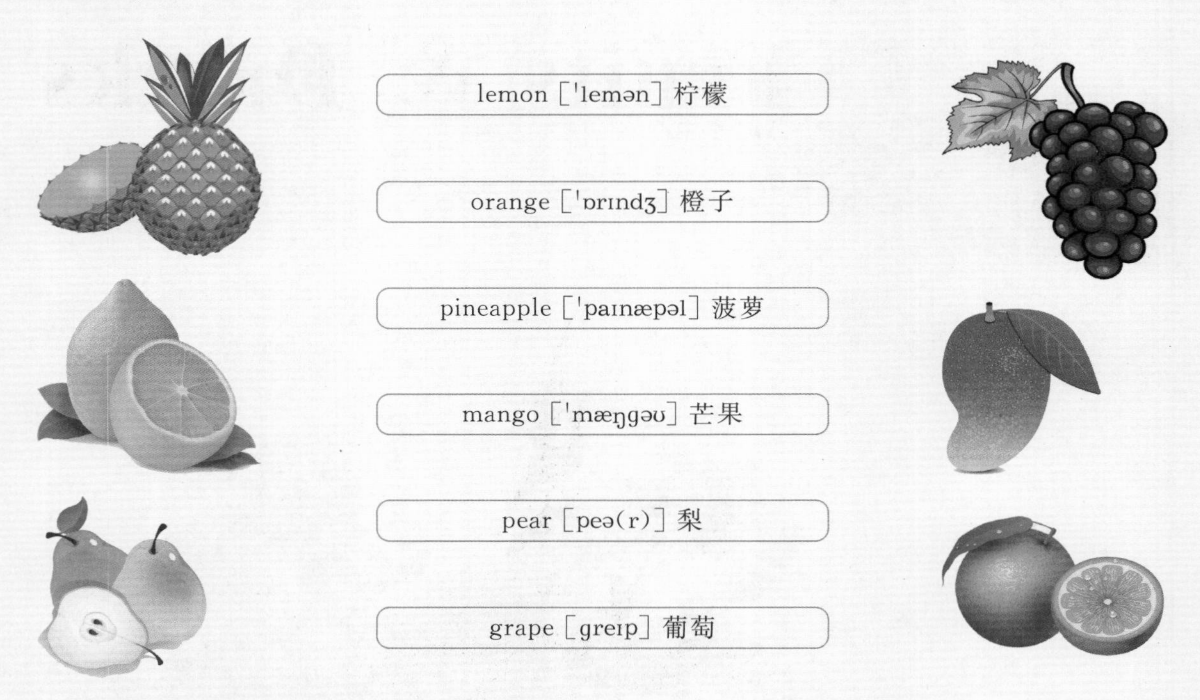

Activity 2 Write. 根据描述写出单词。

1. ________ It's a bright yellow fruit with very sour juice.
2. ________ It's a yellow sweet juicy fruit, which is narrow at the end and wide at the other end.
3. ________ It's purple and grows in bunches (一串) on vines (藤).
4. ________ It has thick brownish skin with sweet, juicy and yellow flesh inside.
5. ________ It's yellow and round, soft and sweet, which has a long hard core.

Student learning(学生学习)

Task 1 Words & expressions 词汇

Task description. 任务描述。

依据构造和特性大致可将水果分为五类：核果类、仁果类、浆果类、橘类及瓜类。本任务将聚焦于水果及常见饮品的单词和词组的学习。

Part 1 Drupe fruits. 核果类水果。

外教有声

 **Activity 1** Read. 读单词和词组。

peach [piːtʃ] 桃子
lychee [ˌlaɪˈtʃiː] 荔枝
red bayberry [ˈbeɪbərɪ] 杨梅
Chinese date 大枣
apricot [ˈeɪprɪkɒt] 杏
plum [plʌm] 梅子
longan [ˈlɒŋgən] 龙眼
cherry [ˈtʃerɪ] 樱桃

Activity 2 Look and write. 看图写中文。

peach ________ apricot ________ lychee ________ plum ________
red bayberry ________ longan ________ Chinesse date ________ cherry ________

Part 2 Pome fruits & berry fruits. 仁果类及浆果类水果。

外教有声

 **Activity 1** Read. 读单词和词组。

haw [hɔː] 山楂
wolfberry [ˈwʊlfbərɪ] 枸杞
strawberry [ˈstrɔːbərɪ] 草莓
blueberry [ˈbluːbərɪ] 蓝莓
loquat [ˈləʊkwɒt] 枇杷
star fruit 杨桃
pomegranate [ˈpɒmɪˌgrænɪt] 石榴
kiwi fruit 猕猴桃

Note

Activity 2 Look and write. 看图写英文。

枇杷________ 山楂________ 蓝莓________ 石榴________

猕猴桃________ 草莓________ 杨桃________ 枸杞________

Part 3 Citrus fruits & melons. 橘类及瓜类水果。

外教有声

Activity 1 Read. 读单词。

grapefruit [ˈgreɪpˌfruːt] 西柚
tangerine [ˌtændʒəˈriːn] 蜜橘
kumquat [ˈkʌmkwɒt] 金橘
melon [ˈmelən] 哈密瓜，甜瓜
pomelo [ˈpɒmɪləʊ] 蜜柚
lime [laɪm] 青柠
papaya [pəˈpaɪə] 木瓜
watermelon [ˈwɔːtəˌmelən] 西瓜

Activity 2 Write and translate. 单词互译。

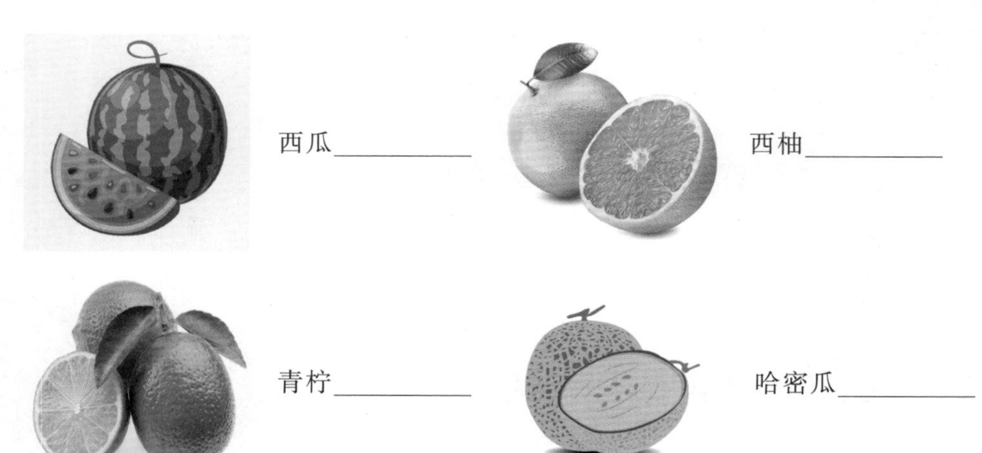

西瓜__________ 西柚__________

青柠__________ 哈密瓜__________

Note

papaya __________

tangerine __________

kumquat __________

pomelo __________

Part 4 Drinks. 饮品。

Herbal tea：花草茶是以花卉植物的花蕾、花瓣或内叶为材料，经过采收、干燥、加工后制作而成的保健饮品。花草茶美容养颜，深受女士们的喜爱。

Activity 1 Read. 读单词和词组。

外教有声

herbal [ˈhɜːbl] tea 花草茶
jasmine [ˈdʒæsmɪn] tea 茉莉花茶
rose tea 玫瑰花茶
vegetable juice 蔬菜汁
green tea 绿茶
black tea 红茶
fruit juice 果汁
milk shake [ʃeɪk] 奶昔

Activity 2 Listen and write. 听录音写单词。

外教有声

____

____

____

____

Note

______ ______ ______ ______

Activity 3 Complete. 补全单词。

1. loq __ __ t 2. pom __ lo 3. t __ nger __ ne 4. h __ w 5. p __ megr __ nate
6. m __ lon 7. p __ __ r 8. w __ lfberry 9. str __ wberry 10. k __ w __ fruit

Activity 4 Making dialogues. 对话练习。

A: What fruit would you like to eat?
B: I'd like to eat ______ .

Task 2 Making & Processing Fruits 水果的加工制作

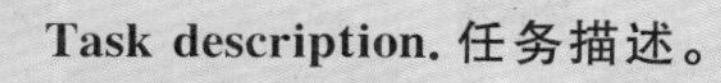

Task description. 任务描述。

水果含有丰富的维生素和膳食纤维等物质，具有极高的营养价值。茶饮具有茶叶的独特风味，是清凉解渴的多功能饮料。任务二将聚焦于果汁等饮品的制作过程的英文表达。

Note

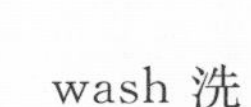

Part 1 Making juice. 果汁的制作。

外教有声

Activity 1 Look and describe. 读单词和词组，用英文描述水果的初加工。

wash 洗　　soak 浸泡　　peel 去皮　　flat 弄平

cut... into half 切成两半　　cut... into cubes 切成方块

remove the seeds from... 去籽　　remove the bottom from... 去蒂

____

____

____

____

____

____

____

____

外教有声

Activity 2 Read. 读单词和词组。

core [kɔː(r)] 果心，核　　do with... 处理

flesh [fleʃ] 果肉　　second ['sekənd] 秒

sound nice 听起来不错

外教有声

Activity 3 Listen and complete. 听录音，补全对话。

A. remove the cores　　B. do with the mangoes

C. add some water　　D. make mango juice

Jack: What shall we __________?

Chef: We'll __________ today.

Jack: What should I do first?

Note

Chef: Wash them and ________.

Jack: I've finished. And then?

Chef: Cut the flesh from mangoes please.

Jack: I see, and put them into the liquidizer?

Chef: Yes, ________ and blend for about 5 seconds. Finally put them into a glass before serving.

Jack: OK, sounds nice!

Activity 4 Write. 写出加工果汁所用工具的英文表达。

____

____

____

Activity 5 Look and choose. 看图并选择。

____

____

____

____

____

A. Put them into a glass before serving.

B. Prepare two mangoes.

C. Add some water.

D. Cut the mangoes and remove cores.

E. Put them into the liquidizer.

Note

Part 2 Making fruit salad. 水果沙拉的制作。

 **Activity 1** Read. 读单词和词组。

外教有声

fashionable ['fæʃnəbl] 时尚的
cherry tomato 圣女果
cut... into pieces 切成小块
dragon fruit ['drægən fruːt] 火龙果
sound amazing [ə'meɪzɪŋ] 听起来非常棒
decorate ['dekəreɪt] with... 用……装饰

Activity 2 Listen and write. 听录音,写出听到的内容。

外教有声

Jack: What shall we make today, chef?
Chef: We are going to ____________.
Jack: OK. What shall we do first?
Chef: ____________. We need bananas, grapes, dragon fruits and some cherry tomatoes.
Jack: Oh, sounds amazing! What's the next?
Chef: Wash them and ____________________, and then cut them into pieces.
Jack: Then put them into the mixing bowl?
Chef: No, this time we use ____________.
Jack: Oh, it looks so fashionable!
Chef: Yes, at last, ____________ on the top.

Activity 3 Look and describe. 看图,参考 **Activity 2** 用英文描述制作过程。

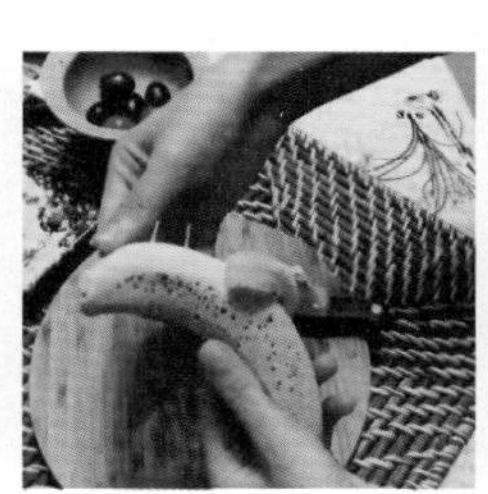

____________ ____________ ____________ ____________

Note

Making other drinks. 其他饮品的制作。

Activity 1 Look and match. 将香蕉奶昔的制作方法与图片连接。

A. Put the bananas into the liquidizer.

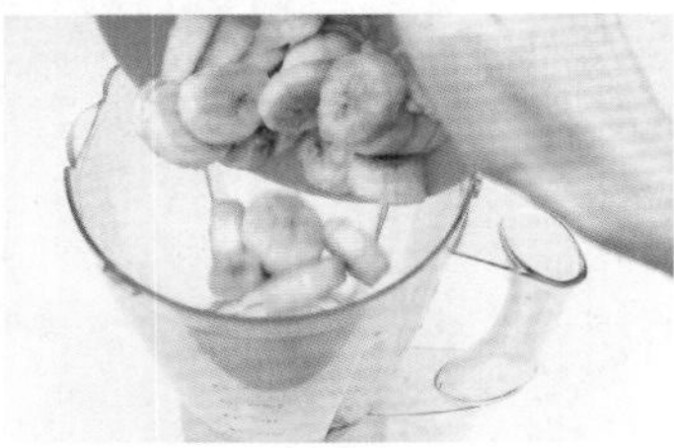

B. Blend for about 5 seconds.

C. Add some milk.

D. Put it into a glass before serving.

E. Prepare the bananas and milk.

Activity 2 Translate into Chinese. 将下列玫瑰花茶泡制过程翻译成中文。

1. You'd better choose a glass tea pot. ____________________

2. Add some rose into the pot. ____________________

3. Pour the hot water into the pot, and make sure that the water temperature is 70 to 80 degrees centigrade(摄氏度).

____

4. It is better to drink with honey.

____

Task 3 Making dishes 菜肴制作

Task description. 任务描述。

水果粥容易消化吸收,大米可提供丰富的B族维生素,具有补中益气、健脾养胃的功效,水果还可以提供各种维生素,营养价值丰富。任务三将聚焦于水果粥制作过程中所需要的原料、调料和加工过程的英文表达。

Note 

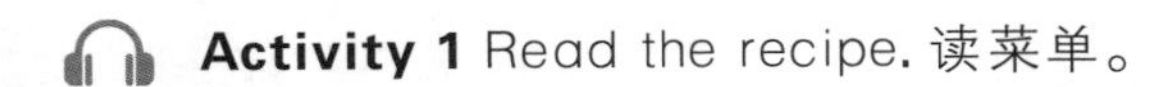

外教有声

Activity 1 Read the recipe. 读菜单。

Fruit congee(水果粥)

Ingredients(原料)

an apple　　a pear　　an orange　　a banana

some longans　　some raisins(葡萄干)　　ice sugar

30 g lotus root starch(藕粉)

Methods(制作方法)

1. Wash the fruits.
2. Peel and cut them into small pieces.
3. Put the lotus root starch into a small bowl, add some cold water and then stir it.
4. Put the fruits into the pot when the water is boiling, and then add some ice sugar.
5. Add the lotus root starch into the pot and stir the congee for a while.
6. Finally, turn off the fire and drop some raisins.

Characteristics(特点)

It has a good taste, rich nutrition(营养) and easy digestion(消化).

Activity 2 Decide true (T) or false (F). 判断正误。

1. ________ Cut the fruits into small pieces when you make the fruits congee.
2. ________ Add some hot water into the lotus root starch.
3. ________ Boil the fruits for about 10 minutes.
4. ________ Add some honey to the pot.
5. ________ Drop some raisins when you turn off the fire.

Activity 3 Translate. 中英文互译。

Ingredients (原料)

香蕉________

苹果________

梨________

橙子________

raisin ________

longan ________

lotus root starch ________

ice sugar ________

Note

Methods (制作方法)

1. Wash the fruits.

____

2. Peel and cut them into small pieces.

____

3. Put the lotus root starch into a small bowl, add some cold water and then stir it.

____

4. Put the fruits into the pot when the water is boiling, and then add some ice sugar.

____

5. Add the lotus root starch into the pot and stir the congee for a while.

____

6. Finally, turn off the fire and drop some raisins before serving.

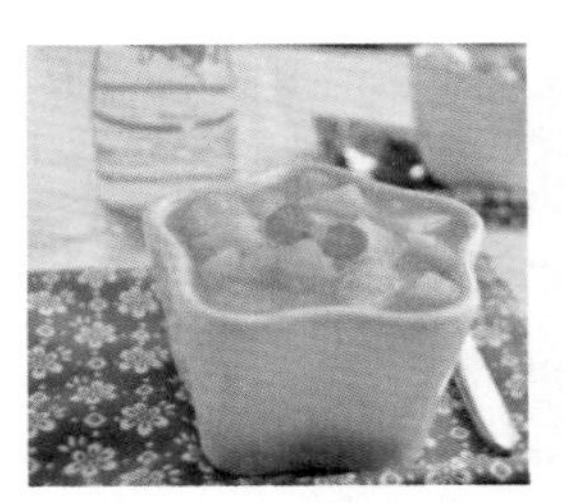

____

Culture knowledge（餐饮文化）

吃水果的学问

色彩斑斓的水果大家一定都喜欢吃，但吃水果也是有学问的，下面我们一起来学习怎样吃水果才健康吧。

❶ **吃菠萝的学问**

菠萝具有健胃消食、解热消暑、解酒、降血压等功效。但属于气管、支气管敏感的人群要注意不可食用，在吃的时候可以先用盐水浸泡，如果食用后感到喉咙不适，可能是过敏症状，需要喝一杯淡盐水稀释。

❷ **吃芒果的学问**

芒果有生津解渴的效果，还具有润泽皮肤、益眼、解毒、降血压等作用。但是肿瘤、过敏、皮肤病人群要谨慎食用。芒果也是哮喘患者的禁忌。

❸ **吃苹果的学问**

苹果中含有果胶、微量元素、有机酸等物质，可以促进胆固醇代谢、降低胆固醇的水平，还有利于高血压患者，对于腹泻也有极好的作用。但溃疡性结肠炎的患者不宜生食苹果，尤其是在急性发作期，由于肠壁溃疡变薄，苹果质地较硬，很不利于肠壁溃疡面的愈合。

❹ **吃香蕉的学问**

香蕉与牛奶同吃可以促进对维生素 B_{12} 的吸收，但西瓜和香蕉同吃会引起腹泻；芋头和香蕉一起吃可能会使胃部不适，感觉胀痛。

❺ **吃荔枝的学问**

荔枝是营养成分极高的水果，但对于一些容易过敏、易上火的人来说，尽量少吃荔枝，否则容易出现头晕、头痛等症状，影响身体健康。

❻ **吃葡萄的学问**

葡萄可以生津消食、缓解疲劳、补血益气，但不能与牛奶同吃。葡萄中含有维生素 C，而牛奶中的有些成分会和葡萄中含有的维生素 C 反应，对胃有伤害，两样同时食用会导致腹泻，严重者还会呕吐。

Unit exercise（单元练习）

Exercise 1 Translate.（单词互译）

1. watermelon ________________　　2. 山楂________________

3. Chinese date ________________　　4. 木瓜________________

5. longan ________________　　6. 荔枝________________

7. loquat ________________　　8. 枸杞________________

9. star fruit ________________　　10. 芒果________________

Exercise 2 Tick off the different words.（找不同）

1. A. peach　　B. apple　　C. Chinese date　　D. apricot

2. A. melon　　B. loquat　　C. grape　　D. pear

3. A. blueberry B. strawberry C. pomegranate D. cherry
4. A. green tea B. fruit juice C. black tea D. jasmine tea

Exercise 3 Match.（连线）

1. What are you doing?
2. What's the color of the grape?
3. Do you like plum?
4. What shall we do with melon?
5. What about the next?

A. Put them in the liquidizer.
B. Just so so, it's too sour.
C. I'm washing the fruits.
D. It's purple.
E. Cut it in half and remove seeds.

Exercise 4 Put words in right order.（句子排序）

1. the, fruits, boil, for, a while

____

2. and, peel,the bananas, cut, then, them, into, pieces

____

3. pineapple, is, color, of, what, the

____

4. shall we, do, What, with, these fruits

____

5. it, seeds, of, then, slice, remove, the, melon

____

Exercise 5 Write.（写出下列饮品的英文表达）

green tea	black tea	herbal tea	jasmine tea
rose tea	fruit juice	vegetable juice	milk shake

红茶__________ 绿茶__________ 果汁__________ 奶昔__________
花草茶__________ 茉莉花茶__________ 玫瑰茶__________ 蔬菜汁__________

Exercise 6 Translate.（将下列水果的加工步骤翻译成英文）

1. 准备好水果。

____

2. 把水果切成小块。

____

3. 放入椰碗中，加入薄荷叶点缀。

____

Exercise 7 Translate.（中英文句子翻译）

1. 今天我们要做草莓奶昔。

____

2. 请把大枣洗干净并浸泡半小时。

____

3. 红茶和绿茶，你要哪一个？

____

4. Add some cold water into the lotus root starch and then stir it.

____

5. Add some rose into the pot and pour into the hot water.

____

Unit 4

参考答案

Unit 5

Vegetables

Learning goal（学习目标）

You will be able to：

1. get familiar with the words and expressions of vegetables；
2. know and write the different shapes of vegetables；
3. describe the steps of making vegetable dishes.

扫码看课件

Lead in(导入)

Activity 1 Look and match. 看图并连线。

snow pea [snəʊ piː] 荷兰豆

onion [ˈʌnjən] 洋葱

leek [liːk] 大葱

eggplant [ˈegplɑːnt] 茄子

tomato [təˈmɑːtəʊ] 番茄

garlic [ˈgɑːlɪk] 蒜

Activity 2 Decide true (T) or false (F). 判断正误。

1. ________ Leeks are green and white and can be used in the porridge(粥).
2. ________ Tomatoes are red and round, and they usually taste sour.
3. ________ The proper amount of garlic can add flavor.
4. ________ Snow peas are round and thin with green color.
5. ________ Onions are suitable for keeping in a dry and dark place.

Student learning(学生学习)

Task 1 Words & expressions 词汇

Task description. 任务描述。

蔬菜是烹饪的重要原料，种类繁多，一般可分成叶类蔬菜、瓜果类蔬菜、花菜类蔬菜、根茎类蔬菜等。其中叶类蔬菜是整个蔬菜中品种最多的类别。任务一将聚焦于各种蔬菜单词和词组的学习。

Note

Part 1 Leafy vegetables. 叶类蔬菜。

外教有声

Activity 1 Read. 读单词和词组。

spinach ['spɪnɪtʃ] 菠菜

cabbage ['kæbɪdʒ] 卷心菜，洋白菜

celery ['seləri] 芹菜

cole [kəʊl] 油菜

parsley ['pɑːslɪ] 香芹，欧芹

spring onion [sprɪŋ 'ʌnjən] 葱

lettuce ['letɪs] 生菜

crown daisy [kraʊn 'deɪzɪ] 茼蒿

Chinese cabbage 大白菜

Activity 2 Look and write. 看图写中文。

lettuce ________ celery ________ cabbage ________

crown daisy ________ spinach ________ cole ________

parsley ________ Chinese cabbage ________ spring onion ________

Note

Part 2 Melon fruit & flower vegetables. 瓜果类及花菜类蔬菜。

外教有声

Activity 1 Read. 读单词和词组。

wax gourd 冬瓜
cucumber ['kju:kʌmbə(r)] 黄瓜
bitter cucumber 苦瓜
broccoli ['brɒkəlɪ] 西蓝花
zucchini [tsʊ:'ki:nɪ] 西葫芦
pumpkin ['pʌmpkɪn] 南瓜
green pepper [gri:n pepə(r)] 青椒
cauliflower ['kɒlɪˌflaʊə(r)] 菜花

Activity 2 Look and write. 看图写英文。

南瓜________

黄瓜________

苦瓜________

西葫芦________

青椒________

冬瓜________

菜花________

西蓝花________

Part 3 Root vegetables. 根茎类蔬菜。

外教有声

Activity 1 Read. 读单词和词组。

turnip ['tə:nɪp] 白萝卜
bamboo shoot [bæm'bu: ʃu:t] 竹笋
carrot ['kærət] 胡萝卜
Chinese yam [jæm] 山药
lotus root ['ləʊtəs ru:t] 藕
asparagus [ə'spærəgəs] 芦笋

Note

Activity 2 Write and translate. 单词互译。

藕________

白萝卜________

山药________

carrot ________

芦笋________

bamboo shoot ________

Activity 3 Complete. 补全单词。

1. __ nion	2. p __ mpkin	3. l __ tt __ ce	4. br __ ccoli	5. c __ bbage
6. t __ rn __ p	7. c __ cumb __ r	8. __ spar __ gus	9. l __ tus r __ ot	10. c __ l __ ry

Activity 4 Making dialogues. 对话练习。

A: What are you cooking?
B: I'm cooking ________.

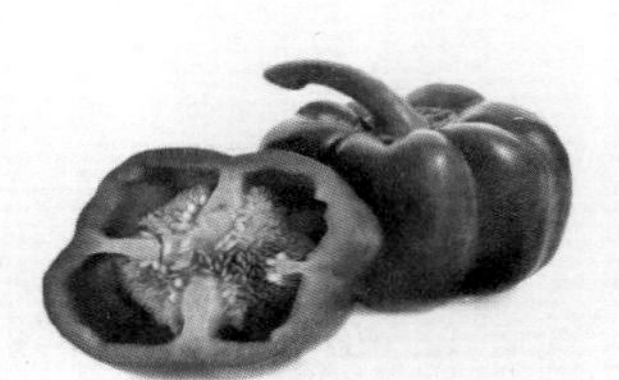

Task 2 Trimming & shaping 初加工和成型

Task description. 任务描述。

蔬菜的初加工是对蔬菜原料在烹调前所进行的选择、整理、洗涤等过程。成型就是根据烹调与食用的需要，将各种蔬菜切成一定形状，使之成为组配菜肴所需要的基本形体的过程。任务二将聚焦于蔬菜的初加工以及成型加工的英文表达。

Part 1 Trimming onions. 洋葱的初加工。

外教有声

Activity 1 Read. 读单词和词组。

layer ['leɪə(r)] 层

skin [skɪn] 外皮

cut ... in half 切成两半

cut up 切碎

slice down through the layers 沿着层切成片

cut few slices but not through the root 向根的方向切片，但不要切透

外教有声

Activity 2 Listen and complete. 听录音，补全对话。

A. slice down through the layers

B. cut the onion in half to peel the skin

C. cut across slices to make dices

D. Cut few slices but not through the root

Chef: Do you know what we are going to do today, Jack?

Jack: Yes, We are going to trim onions.

Chef: Right. What is the first step?

Jack: Um, shall I ____________, chef?

Chef: Yes, Jack. Go ahead.

Jack: I've done it. What's the next?

Chef: ____________.

Jack: OK, not through the root. Shall I dice it now?

Note

Chef: No, ____________.

Jack: OK, I'll do it, chef.

Chef: Good. Last ____________.

Jack: I'll do more to be a good cook.

Chef: No problem, believe yourself.

Activity 3 Choose and write. 选出洋葱初加工所用工具并写出英文表达。

____

____

____

Activity 4 Look and choose. 看图并选择。

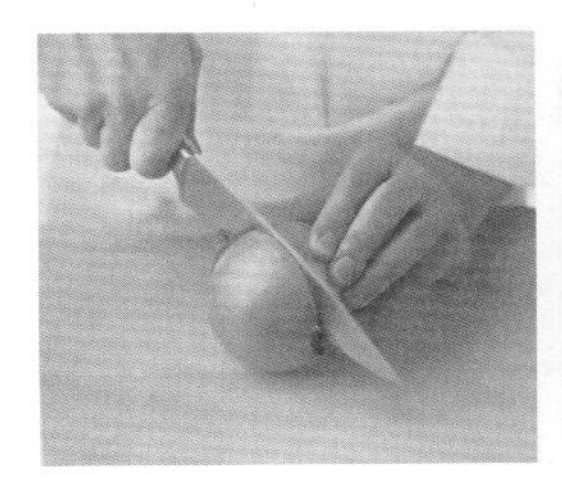

____

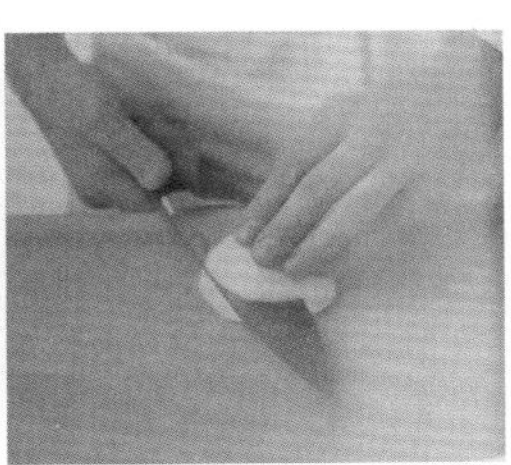

____

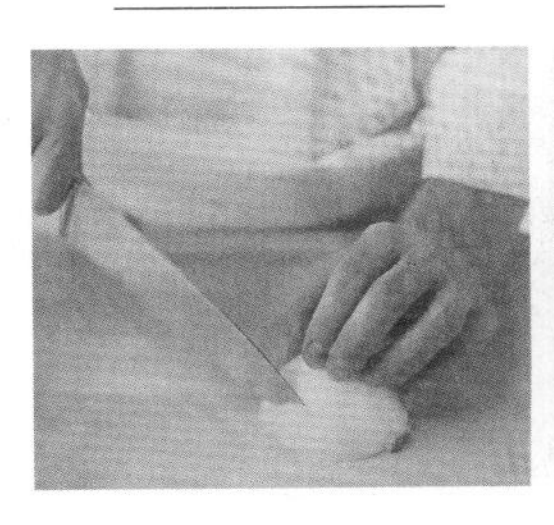

____

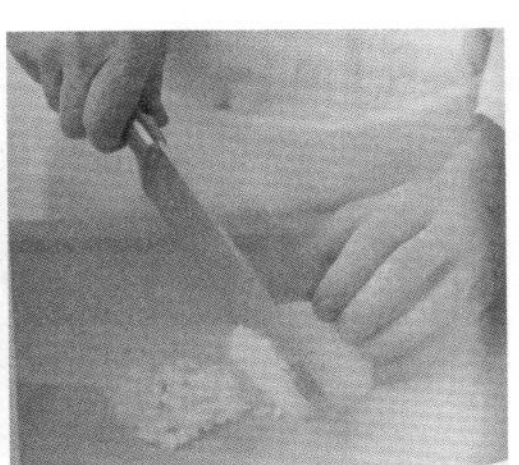

____

A. Slice down through the layers.

B. Cut the onion in half to peel the skin.

C. Cut across slices to make dices.

D. Cut few slices but not through the root.

Part 2 Trimming bell peppers. 灯笼椒的初加工。

Activity 1 Read. 读单词和词组。

外教有声

bottom [ˈbɒtəm] 底部，根蒂

slice [slaɪs] 切薄片，划

bell pepper 灯笼椒

slice in half 切成两半

cut off the top 切掉蒂

rib [rɪb] 肋条，筋

section [ˈsekʃən] 部分，块

according to 根据

Note

外教有声

Activity 2 Listen and write. 听录音，写出听到的内容。

Jack: I've learned the way of trimming onions. What shall I do with the bell peppers, chef?

Chef: You should ____________ first.

Jack: OK, it's so easy.

Chef: Secondly, remove the ________.

Jack: It needs the skill, right?

Chef: Yes, and then cut the pepper into ________.

Jack: I've done it.

Chef: So trim the sections ____________ your dish.

Jack: Yes, I know that, chef.

Activity 3 Look and describe. 看图并参考 **Activity 2** 用英文描述。

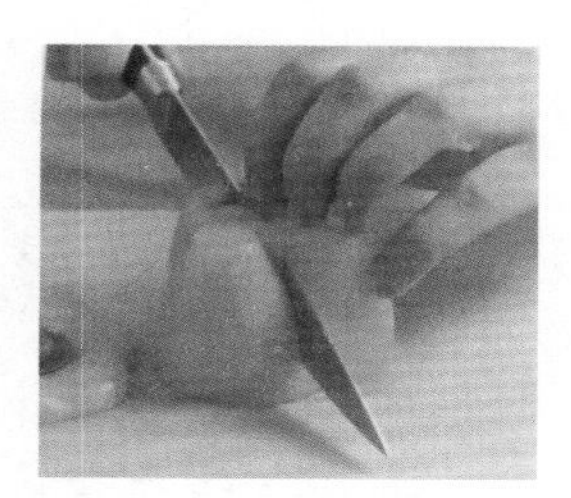

1. ____________________________

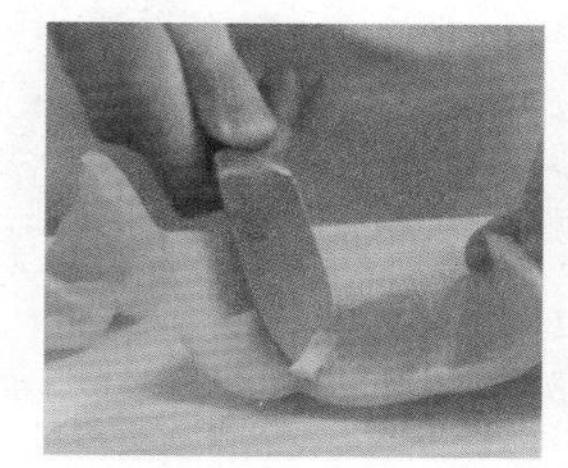

2. ____________________________

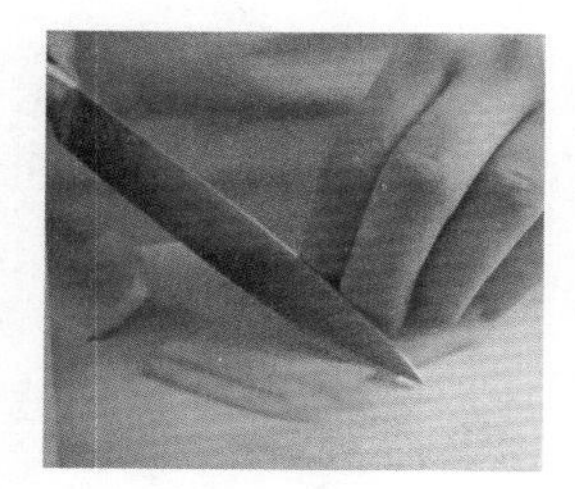

3. ____________________________

Note

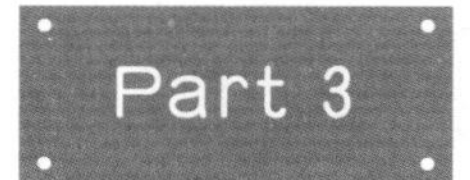

Part 3 Trimming other vegetables. 其他蔬菜的初加工。

Activity 1 Look and match. 将紫甘蓝的去心和切丝方法与图片连接。

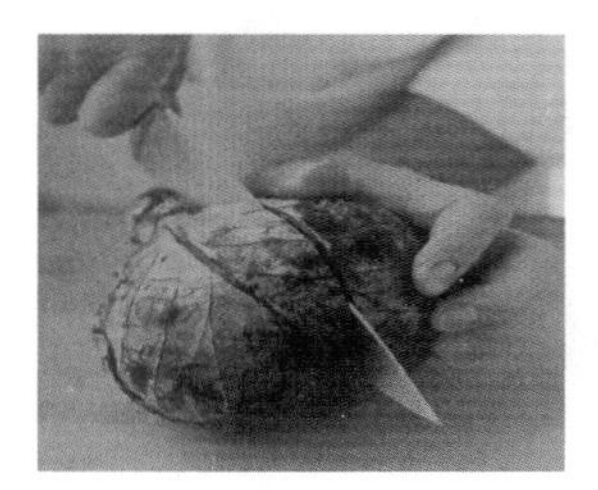

A. Slice out the core from each quarter.

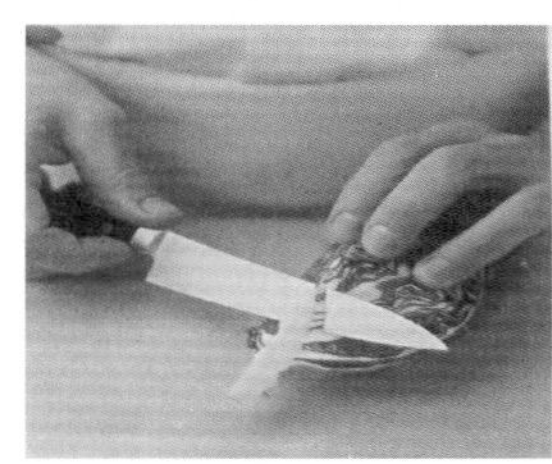

B. Cut across the purple cabbage with shreds.

C. Cut the purple cabbage in half.

Activity 2 Translate into Chinese. 将下列句子翻译成中文。

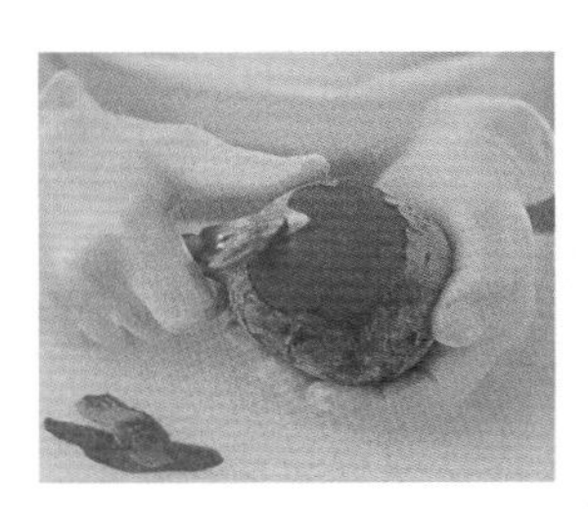

1. Peel the vegetable with a peeler.

____

2. Trim the sides to form a square shape.

____

Note

3. Cut it into equal slices.

____

4. Cut them into square -edged batons.

____

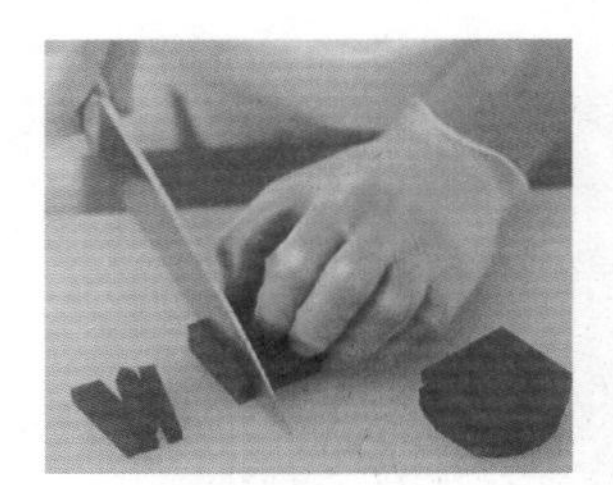

Task 3 Making dishes 菜肴制作

Task description. 任务描述。

荷兰豆的营养价值很高，它富含维生素C和能分解体内亚硝胺的酶，可以分解亚硝胺，且有较为丰富的膳食纤维，具有防癌抗癌的作用。任务三将聚焦于蒜蓉荷兰豆制作过程中所需要的原料、调料和加工过程的英文表达。

外教有声

Activity 1 Read the recipe. 读菜单。

Snow peas with grinded garlic(蒜蓉荷兰豆)

Ingredients(原料)

45 mL peanut oil　　200 g snow peas

1/2 tbsp of soy sauce　　2 tbsp of grinded garlic

1/4 tsp of gourmet powder　　1/4 tsp of salt

Methods(制作方法)

1. Boil the snow peas in boiling water for two minutes and dry them.
2. Stir-fry the grinded garlic in hot peanut oil for two minutes until heated through.
3. Put the snow peas into peanut oil, stir-fry for two minutes.
4. Add soy sauce, salt and gourmet powder, stir-fry for one minute.

Characteristics(特点)

It is crisp and fresh, with an attractive green color.

Activity 2 Decide true (T) or false (F). 判断正误。

1. ________ Boil snow peas in cold water for two minutes.
2. ________ Stir-fry snow peas in peanut oil first.
3. ________ Stir-fry grinded garlic in oil for one minute.
4. ________ Snow peas with grinded garlic is sweat and green.
5. ________ We don't need soy sauce while cooking this dish.

Activity 3 Translate. 中英文互译。

Ingredients(原料)

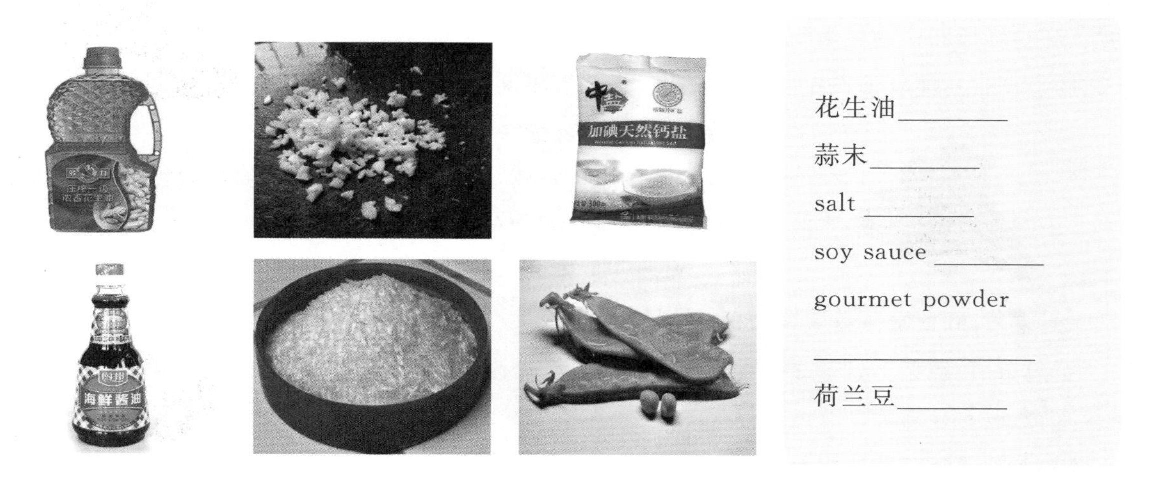

花生油________

蒜末________

salt ________

soy sauce ________

gourmet powder

____

荷兰豆________

Methods(制作方法)

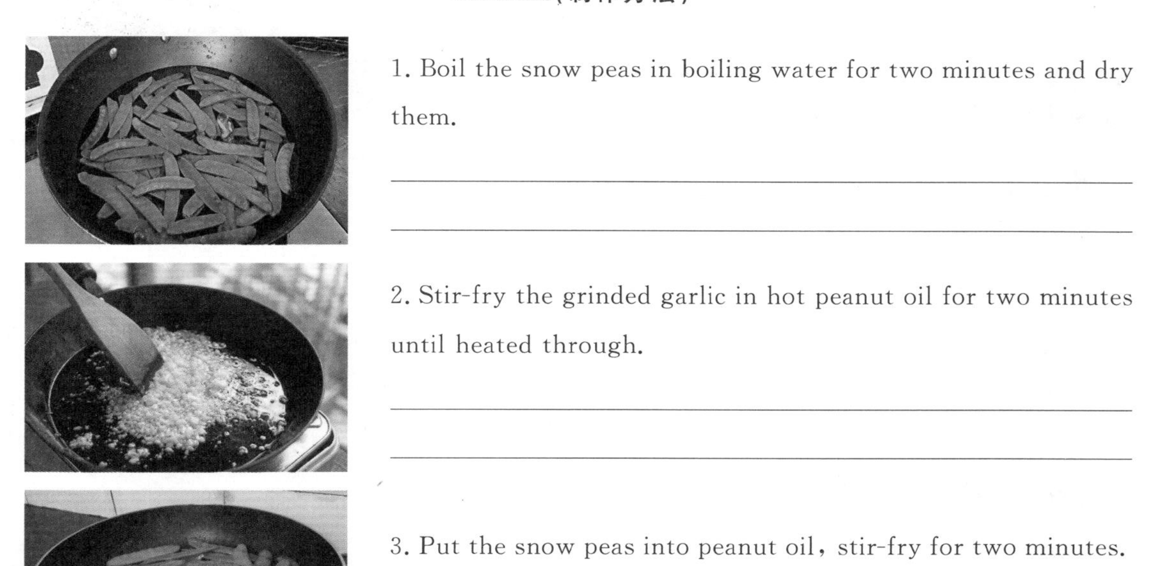

1. Boil the snow peas in boiling water for two minutes and dry them.

____

____

2. Stir-fry the grinded garlic in hot peanut oil for two minutes until heated through.

____

____

3. Put the snow peas into peanut oil, stir-fry for two minutes.

____

____

4. Add soy sauce, salt and gourmet powder, stir-fry for one minute.

____

____

Culture knowledge(餐饮文化)

不同颜色的蔬菜营养价值有何不同?

不同颜色的蔬菜有蔬菜共同的营养价值,即普遍含有维生素、矿物质、膳食纤维和植物化学物,对于满足人体微量营养素的需要、保持人体肠道正常功能有重要作用。但它们各自又有不同的营养特点。

❶ 绿色蔬菜叶酸多

绿色蔬菜含有丰富的叶酸,叶酸对胎儿的作用极其重要,同时绿色蔬菜也是钙元素的优质来源,而且这类蔬菜还含有比较多的维生素 C、类胡萝卜素、铁和硒等微量元素。

❷ 黄色、红色蔬菜改善夜盲症

黄色、红色蔬菜富含胡萝卜素和维生素 C,能提高食欲、刺激神经系统兴奋、改善夜盲症。

❸ 多吃紫色蔬菜让你美丽又年轻

紫色蔬菜富含花青素,具有很强的抗氧化作用,能预防心脑血管疾病,可提高机体的免疫力。

Unit exercise(单元练习)

Exercise 1 Translate.(单词互译)

1. pumpkin ________________ 2. 西蓝花________________
3. spinach ________________ 4. 青椒________________
5. eggplant ________________ 6. 大白菜________________
7. asparagus ________________ 8. 茼蒿________________
9. bell pepper ________________ 10. 荷兰豆________________

Exercise 2 Tick off the different words.(找不同)

1. A. eggplant B. potato C. onion D. lettuce
2. A. bean B. pea C. green pepper D. parsley
3. A. pumpkin B. cabbage C. garlic D. cucumber
4. A. asparagus B. lettuce C. carrot D. turnip

5. A. lettuce B. tomato C. cabbage D. celery

Exercise 3 Fill in the blanks.(选择合适的单词填空)

peel grind chopping put... to batons sliced

1. ________ the potatoes firstly.

2. For the preparation work, ________ the garlic finely.

3. Please ________ some salt ________ the broccoli.

4. Would you cut some ________ onion for black fungus with onion?

5. —Shall I chop the carrots?

—No, cut carrot ________ carefully.

6. —What are you doing?

—I'm ________ the celery.

Exercise 4 Put words in right order.(句子排序)

1. is, the, most, cold dishes, of, onion, "heart", of

____

2. slicing, green pepper, are, now, you

____

3. I, shall, dice, yam, the

____

4. for, chop, celery, the, dish, the, please

____

5. cook, I, shall, how, spinach, the

____

Exercise 5 Write.(写出下列蔬菜加工的英文表达)

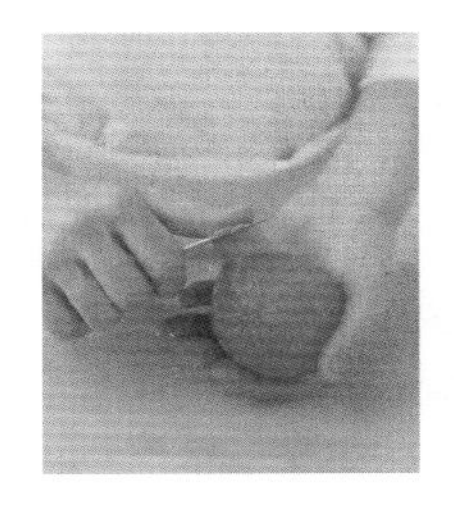

1. ____________

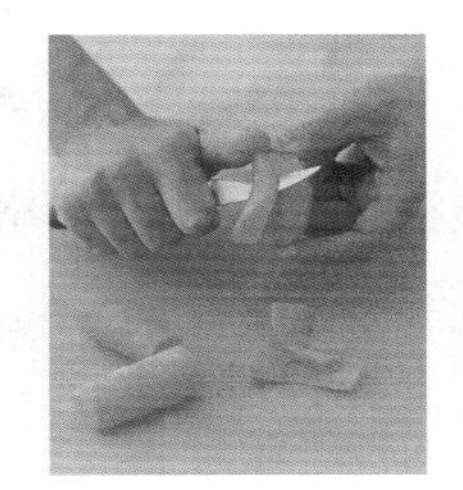

2. ____________

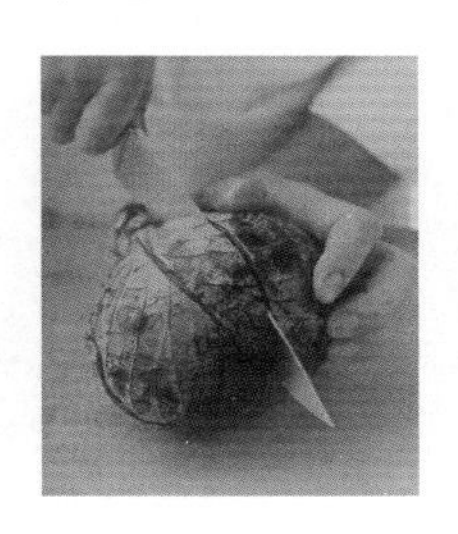

3. ____________

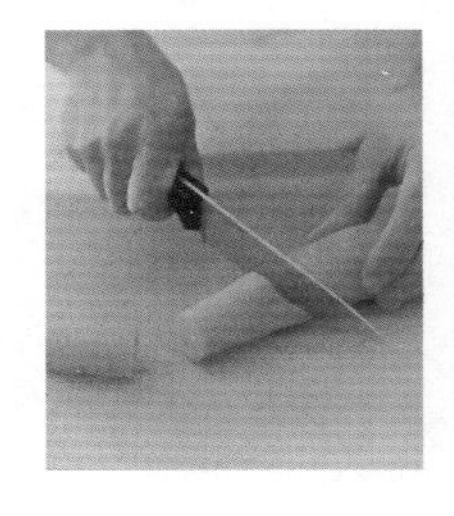

4. ____________

1. 去番茄皮。
2. 把胡萝卜两边去掉。
3. 把圆白菜(紫甘蓝)直切成两半。
4. 把胡萝卜切成段。

Exercise 6 Translate.（将洋葱的初加工方法翻译成中文）

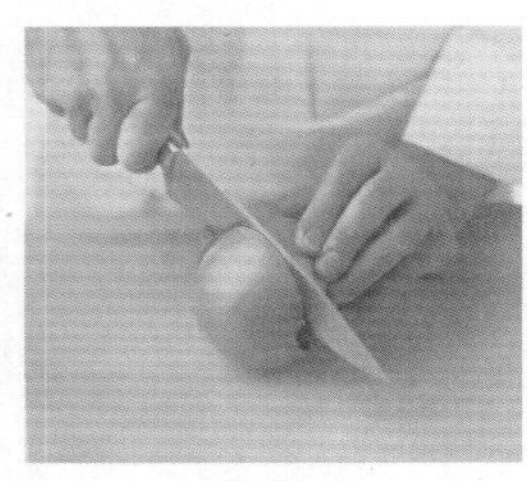

1. 切成两半，去皮。

____

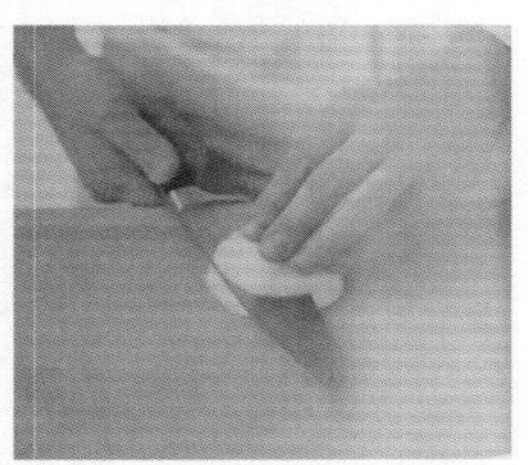

2. 切洋葱片。

____

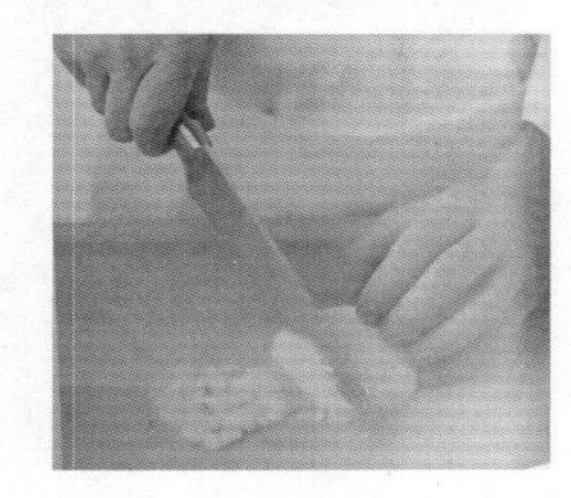

3. 切洋葱丁。

____

Exercise 7 Translate.（中英文句子翻译）

1. 我们需要切得很细的洋葱丝。

____

2. 我现在切白萝卜片好吗？

____

3. 他正在剥蒜。

____

4. Please wash the pumpkin, and then dice it.

____

5. Salt the cucumber in five minutes.

____

Unit 5

参考答案

Note

Unit 6

Meat & Poultry

Learning goal（学习目标）

You will be able to：

1. know the words and expressions of meat and poultry；
2. get familiar with the cutting and cooking ways；
3. talk about the steps of making meat dishes.

扫码看课件

Lead in（导入）

Activity 1 Look and match. 看图并连线。

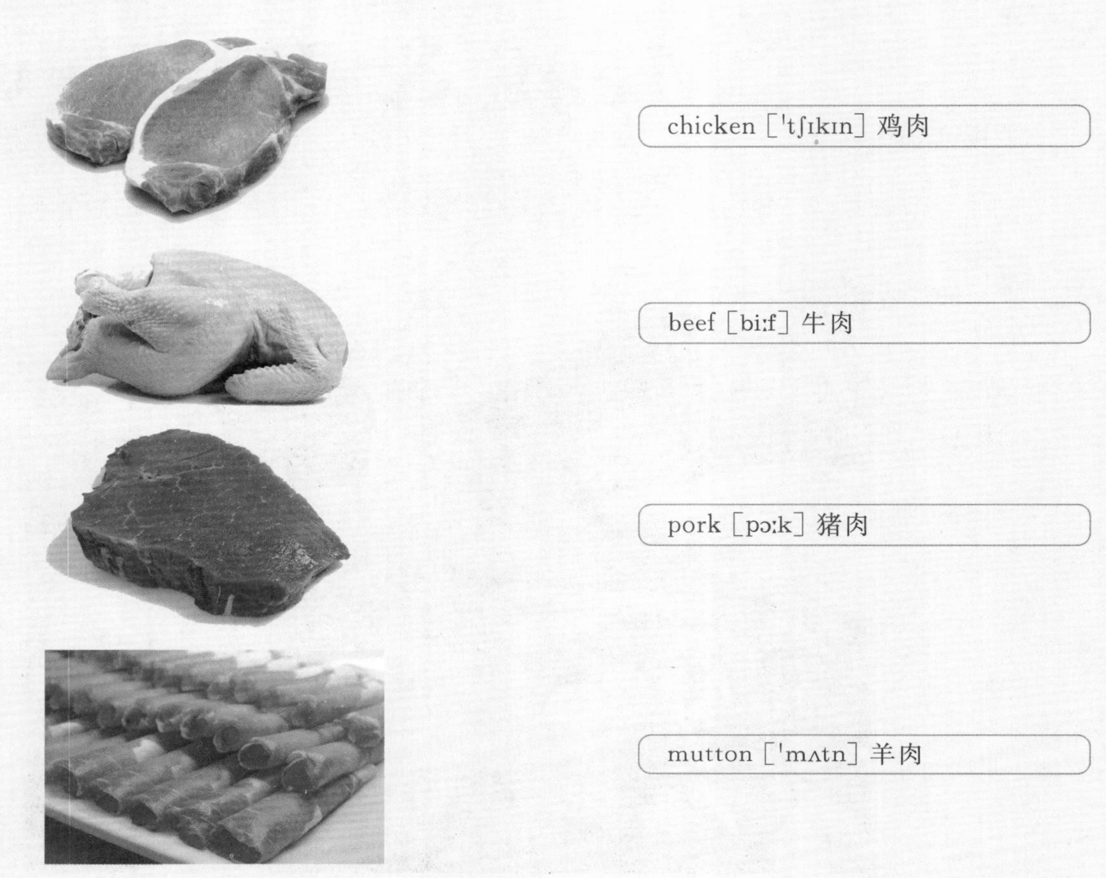

Activity 2 Decide true (T) or false (F). 判断正误。

1. ________ Pork has more fat than beef and mutton.
2. ________ Chicken belongs to poultry.
3. ________ We should eat meat as much as possible every day.
4. ________ People in Muslim countries can't eat pork.
5. ________ Fried food is delicious, so you can eat too much or too often.

Student learning(学生学习)

Task 1 Words & expressions 词汇

Task description. 任务描述。

肉是日常生活的主要副食品,含有丰富的蛋白质及脂肪、碳水化合物、钙、铁、磷等营养成分,具有补虚强身、滋阴润燥、丰肌泽肤的作用。任务一将聚焦于各种肉类单词和词组的学习。

Part 1 Pork. 猪肉。

Activity 1 Read. 读单词和词组。

外教有声

pork shoulder [ˈʃəʊldə(r)] 肩胛肉
pork butt [bʌt] 梅花肉
hind [haɪnd] feet 蹄髈
pork shoulder feet 前腿肉
pork tenderloin [ˈtendəlɔɪn] 小里脊
pork belly [ˈbelɪ] 五花肉
sparerib [ˈspeərɪb] 排骨
pork ham [hæm] 后腿肉
pork loin [lɔɪn] 大里脊

Activity 2 Look and write. 看图写中文。

pork belly ________

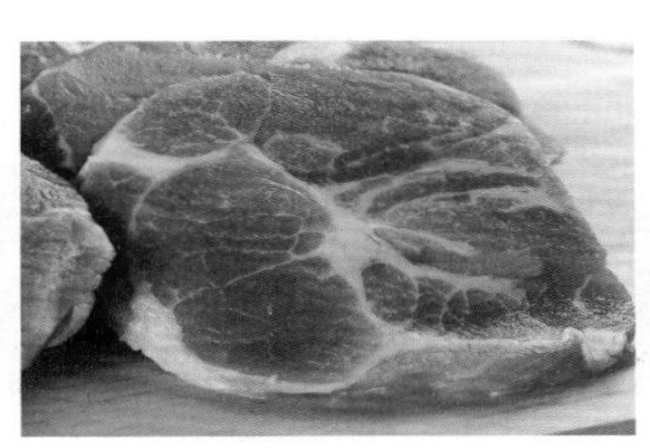

pork butt ________

spareribs ________

pork loin ________

pork tenderloin ________

hind feet ________

Note

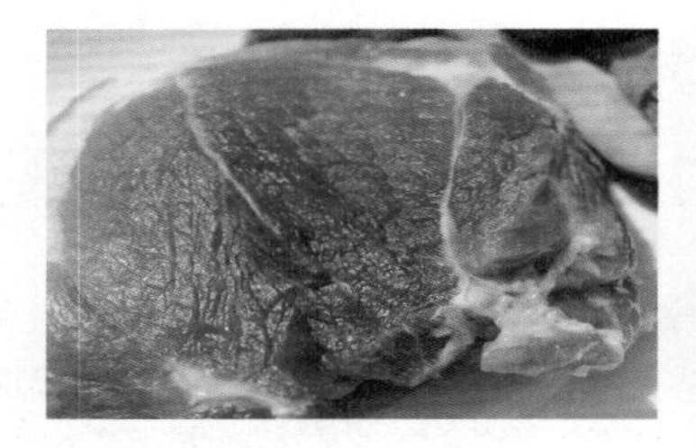

pork shoulder feet ________

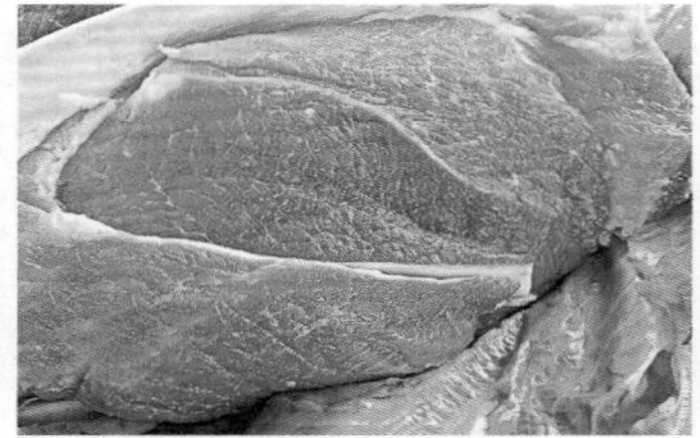

pork ham ________

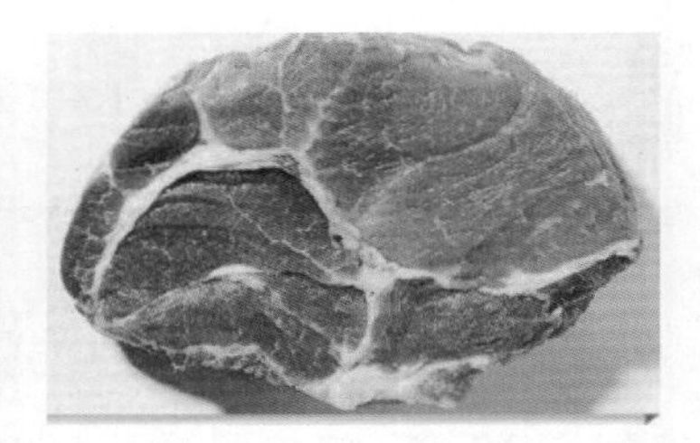

pork shoulder ________

Part 2 Beef & lamb. 牛肉和羊肉。

外教有声

Activity 1 Read. 读单词和词组。

chuck [tʃʌk] 牛肩肉

topside [ˈtɒpsaɪd] 米龙（臀肉）

sirloin [ˈsɜːlɔɪn] 牛腩

lamb rack [ræk] 羊肋排

lamb brisket [ˈbrɪskɪt] 羊胸肉

rib eye [rɪb aɪ] 牛眼肉

beef shank [ʃæŋk] 牛腱

lamb leg [læm leg] 羊腿

lamb fillet [ˈfɪlɪt] 羊上脑

Activity 2 Look and write. 看图写英文。

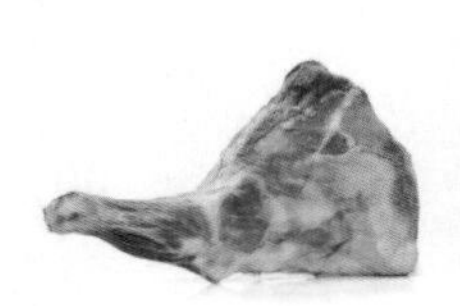

羊腿________

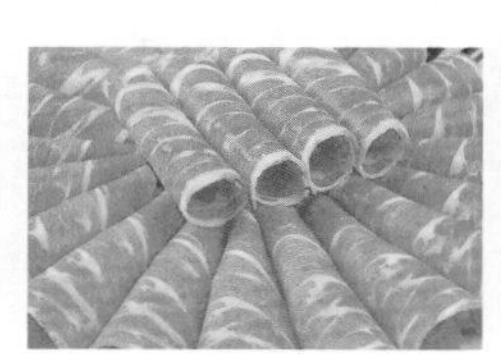

羊上脑________

羊胸肉________

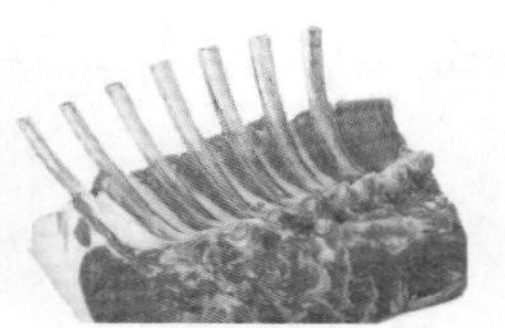

羊肋排________

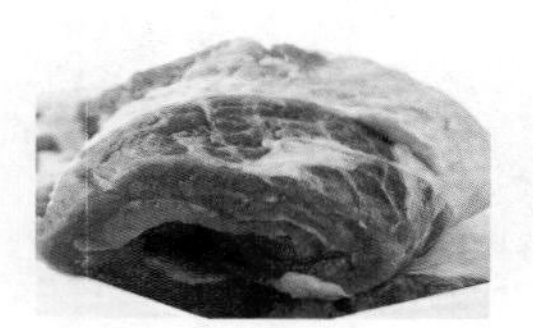

牛腩________

牛腱________

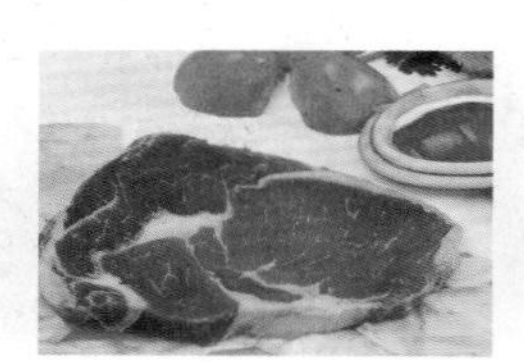

牛眼肉________

牛肩肉________

Part 3 Chicken. 鸡肉。

Activity 1 Read. 读词组。

chicken breast [brest] 鸡胸
chicken wing [wɪŋ] 鸡翅
chicken neck [nek] 鸡脖
chicken liver [ˈlɪvə(r)] 鸡肝
chicken drumstick [ˈdrʌmstɪk] 鸡腿
chicken gizzard [ˈɡɪzəd] 鸡胗
chicken claw [klɔː] 鸡爪
chicken heart [hɑːt] 鸡心

Activity 2 Translate and write. 单词互译。

鸡腿 ________________

鸡胸 ________________

鸡翅 ________________

鸡胗 ________________

chicken claw ________________

chicken neck ________________

chicken heart ________________

chicken liver ________________

Activity 3 Complete. 补全单词。

1. sp __ __ __ ribs 2. sh __ nk 3. eye r __ bs 4. br __ __ st 5. l __ mb l __ g
6. t __ ps __ de 7. l __ ve __ 8. ch __ ck 9. s __ rl __ __ n 10. b __ lly

Activity 4 Making dialogues. 对话练习。

A: Which part of meat do we usually fry for a meal?
B: We usually fry ________ .

Task 2 Trimming meat 肉的初加工

Task description. 任务描述。

中餐中常见的肉类原料有猪肉、羊肉和鸡肉等，形式上又有鲜肉和冻肉两类。由于烹饪技法的不同，初步处理的方法也不尽相同。本任务将聚焦于刀工和烹饪技法、羊肉和鸡肉初加工的单词、句型以及加工成型的英文表达。

Part 1 Cutting & cooking ways. 刀工和烹饪技法。

外教有声

Activity 1 Read. 读单词。

slice [slaɪs] 肉片

shred [ʃred] 肉丝

block [blɒk] 大块肉

cube [kjuːb] 四方块肉

dice [daɪs] 切块

mince [mɪns] 肉馅

Activity 2 Listen and complete. 听录音，补全对话。

外教有声

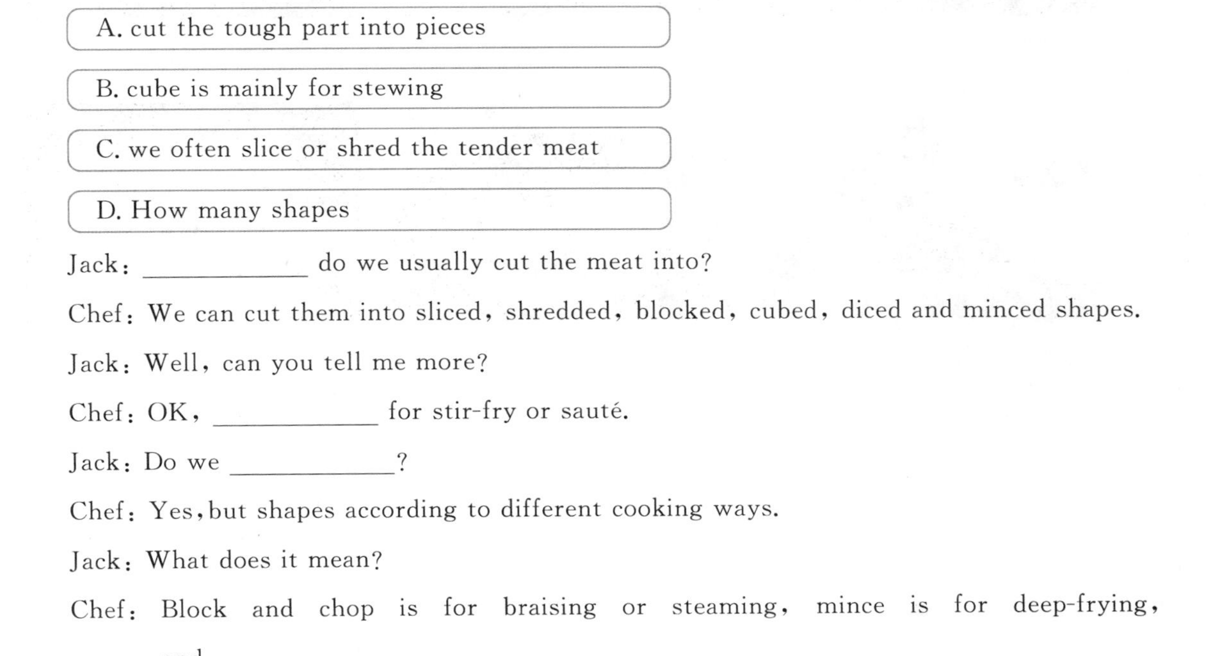

A. cut the tough part into pieces

B. cube is mainly for stewing

C. we often slice or shred the tender meat

D. How many shapes

Jack: ____________ do we usually cut the meat into?

Chef: We can cut them into sliced, shredded, blocked, cubed, diced and minced shapes.

Jack: Well, can you tell me more?

Chef: OK, ____________ for stir-fry or sauté.

Jack: Do we ____________?

Chef: Yes, but shapes according to different cooking ways.

Jack: What does it mean?

Chef: Block and chop is for braising or steaming, mince is for deep-frying, and ____________.

Jack: Sounds interesting! I remember that.

Note

Activity 3 Look and write. 看图并写出下列形状的英文表达。

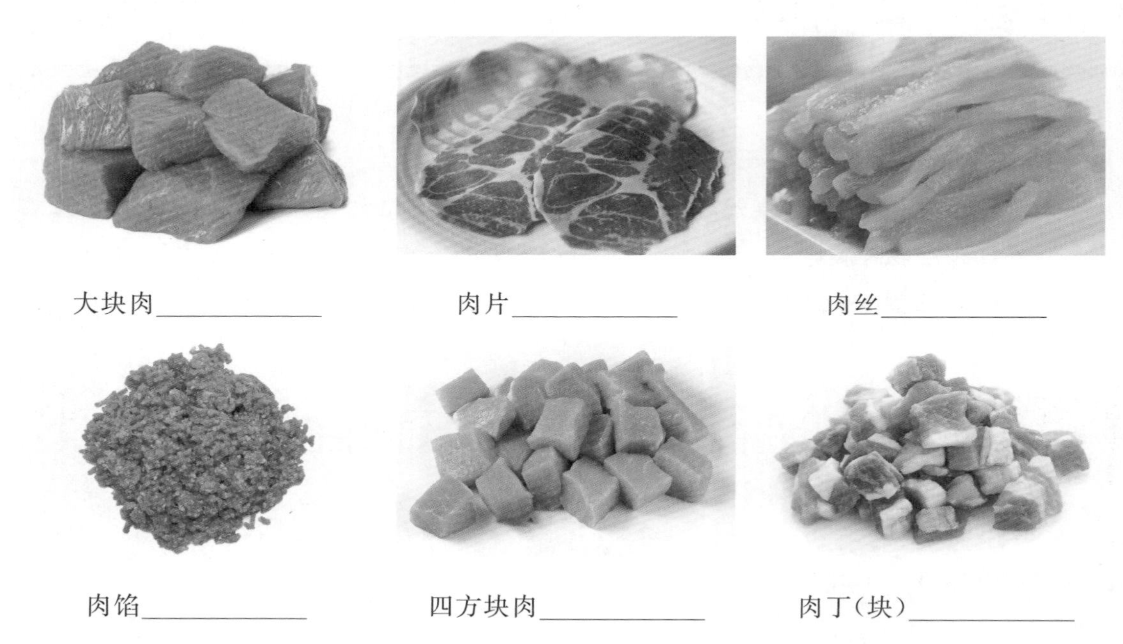

大块肉______ 肉片______ 肉丝______

肉馅______ 四方块肉______ 肉丁(块)______

Activity 4 Write. 参考 **Activity 2** 用英文描述烹饪技法。

炖______ 嫩煎______ 蒸______

煨______ 翻炒______ 深炸______

Note

Part 2 Trimming lamb. 羊肉的初加工。

外教有声

Activity 1 Read. 读单词和词组。

butterfly ['bʌtəflaɪ] 分割成蝴蝶状
sinew ['sɪnjuː] 肌肉
long-bladed ['bleɪdɪd] knife 长柄刀
pelvis ['pelvɪs] 盆骨
tendon ['tendən] 腱
the ball and socket joint 球窝关节

外教有声

Activity 2 Listen and write. 听录音，写出听到的内容。

Jack: How to butterfly a leg of lamb?

Chef: We usually cut around the leg bone with a sharp knife.

Jack: Shall I use a ________________?

Chef: Yes, and then cut ________________.

Jack: What shall I do next?

Chef: Cut away the flesh from ________________.

Jack: I've finished. It's skillful.

Chef: OK, now cut through ________________.

Jack: And then?

Chef: ________________ the thick meaty pieces on either side.

Jack: Thank you, chef. I learned a lot today.

Chef: You're welcome.

Activity 3 Write. 写出加工羊肉所用工具的英文表达。

____

Activity 4 Look and choose. 看图并选择。

A. Cut away the flesh from the ball and socket joint.

B. Cut through the thick meaty pieces.

C. Cut around the leg bone with a sharp knife.

Note

D. Cut through the sinew and tendons.

E. Cut from the pelvis to the bottom.

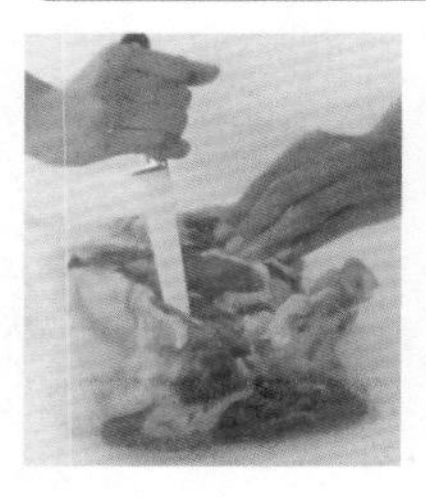 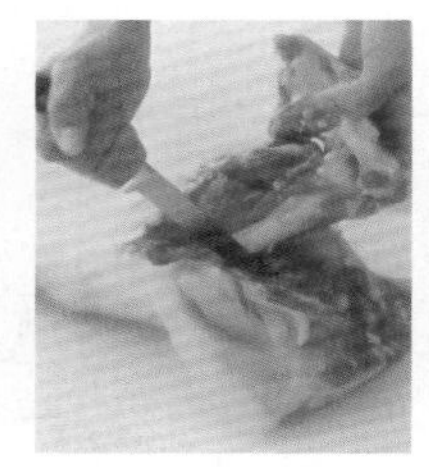   

________ ________ ________ ________ ________

Part 3 Trimming chicken. 鸡肉的初加工。

Activity 1 Look and match. 将加工鸡胸的方法与图片连接。

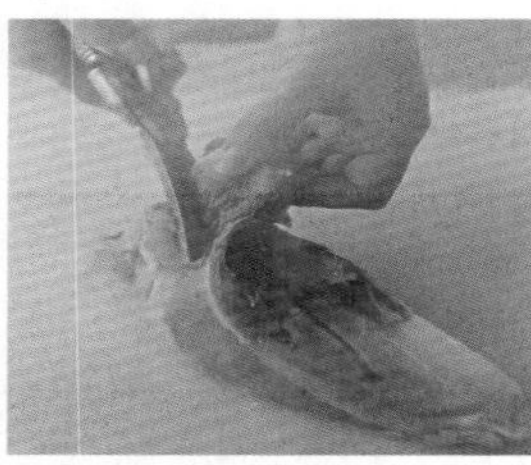

A. Remove the fillet and cut away any connecting membrane.

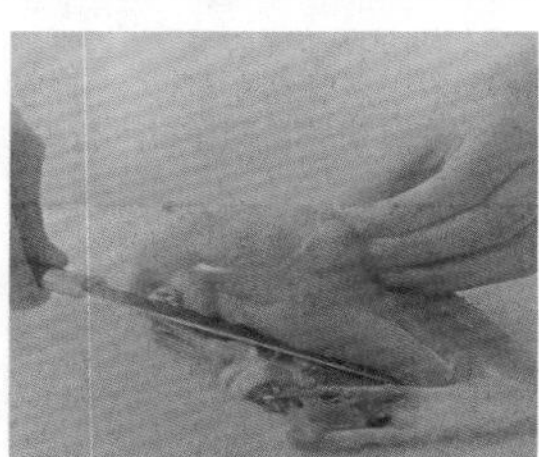

B. Cut away the backbone and ribs with sharp poultry shears.

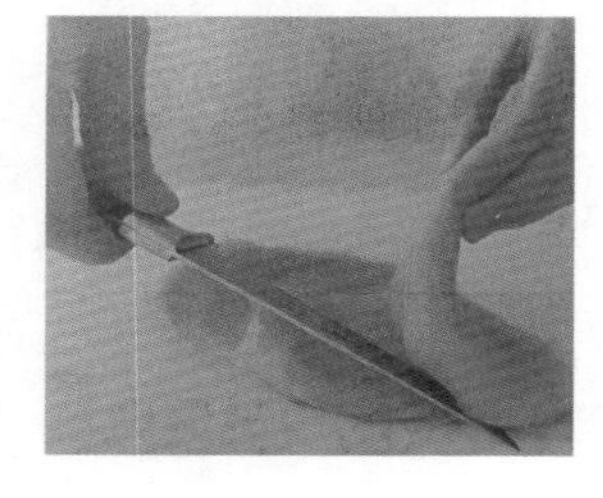

C. Cut the meat away from the bone with a sharp knife.

Activity 2 Translate into Chinese. 将剔除鸡腿骨的方法翻译成中文。

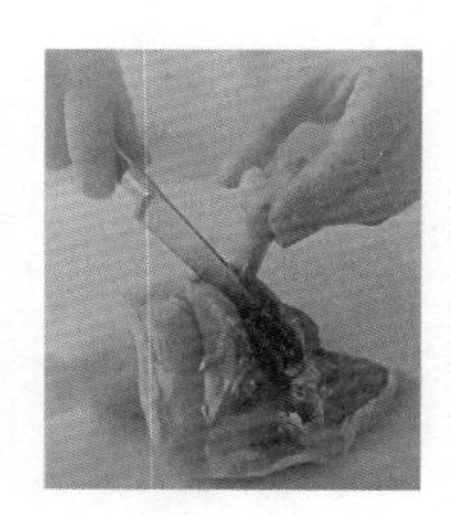

1. Cut the flesh away from the start of the thigh bone.

____

Note

2. Cut all the way down the length of the chicken drumstick.

____

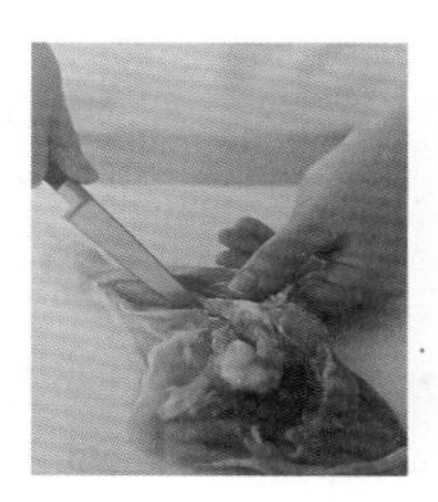

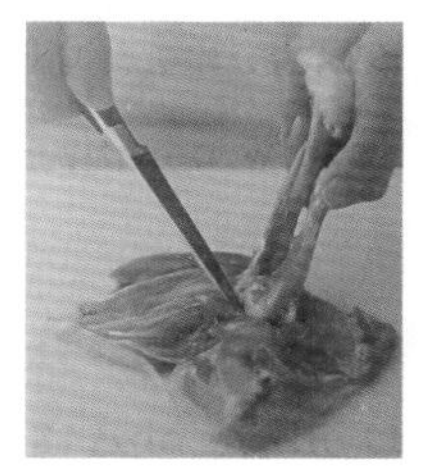

3. Lift the bones up from the central knuckle joint.

____

Task 3 Making dishes 菜肴制作

Task description. 任务描述。

在烹饪肉类菜肴时，不仅要注意保留食物中的营养元素，而且应合理搭配一些蔬菜、水果等碱性食物，这样可以中和酸性，防止营养代谢性疾病的发生。任务三将聚焦于宫保鸡丁制作过程中所需要的原料、调料和加工过程的英文表达。

外教有声

Activity 1 Read the recipe. 读菜单。

Spicy diced chicken with peanuts(kung pao chicken，宫保鸡丁)

Ingredients(原料)

chicken breast　fried peanuts　chopped scallion　sliced ginger　crushed garlic　salt
dried red chili　soy sauce　sugar　vinegar　water starch　white pepper
cooking wine

Methods (制作方法)

1. Make seasoning sauce with soy sauce, sugar, sliced ginger and white pepper.

2. Put the edible oil for frying seven-eight into hot until the diced chicken breast fried white and take away the heat.

3. Add dried red chili, chopped scallion, sliced ginger and white pepper to stir flavor.

4. Put fried chicken breast and seasoning sauce for stir-fry.

5. Turn off the heat, stir in fried peanuts and serve it on the table.

Characteristics(特点)

It is tender while the peanuts are crisp, which is spicy and tasty.

Activity 2 Decide true (T) or false (F). 判断正误。

1. ________ Spicy diced chicken breast with peanuts is welcomed by foreigners.
2. ________ We should fry the diced chicken breast for a very long time.
3. ________ Peanuts are necessary to make the dish crispy.
4. ________ Both the ginger and scallion make the dish very spicy.
5. ________ We can fry breast drumstick instead of chicken breast.

Activity 3 Translate. 中英文互译。

Ingredients (原料)

chicken breast ____________
fried peanuts ____________
sliced ginger ____________
cooking wine ____________
crushed garlic ____________
chopped scallion ____________
干辣椒____________
酱油____________
淀粉____________
白胡椒____________
糖____________
醋____________

Methods (制作方法)

1. Put the edible oil for frying seven-eight into hot until the diced chicken breast fried white and take away the heat.

____

2. Add dried red chili, chopped scallion, sliced ginger and white pepper to stir flavor.

____

3. Put fried chicken breast and seasoning sauce for stir-fry.

____

4. Turn off the heat, stir in fried peanuts and serve it on the table.

____

Culture knowledge（餐饮文化）

中餐有哪些烹饪方法?

❶ **煎**　一般日常所说的煎，是指用锅把少量的油加热，再把食物放进去，使其熟透。食物表面会稍成金黄色乃至微煳，煎出来的食物味道也会比水煮的甘香可口。例如煎饺、煎豆腐、韭菜盒子、煎饼等。

❷ **炒**　炒是最基本的烹饪技法，其原料一般是片、丝、丁、条、块。炒时要用旺火，要热锅热油，所用底油多少随料而定。依照材料、火候、油温高低的不同，炒可分为生炒、滑炒、熟炒及干炒等方法。

❸ **蒸**　蒸是指把经过调味后的原料放在器皿中，再置入蒸笼利用蒸汽使其成熟的过程，可分为猛火蒸、中火蒸和慢火蒸三种。例如粉蒸肉、清蒸螃蟹、清蒸武昌鱼、蒸水蛋等。

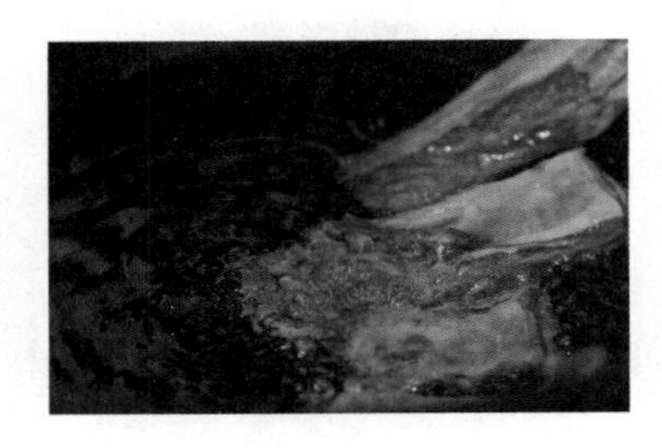

❹ **炸** 炸是一种旺火、多油、无汁的烹饪方法。炸有很多种，如清炸、干炸、软炸、酥炸、面包渣炸、纸包炸、脆炸、油浸、油淋等。

❺ **焖** 焖是将锅置于微火上加锅盖把菜做熟的一种烹饪方法。其操作过程与烧很相似，但小火加热的时间更长，一般在半小时以上，火力也更小。

Unit exercise(单元练习)

Exercise 1 Translate.(单词互译)

1. stir-fry ____________ 2. 片____________
3. cube ____________ 4. 蒸____________
5. pork belly ____________ 6. 煨、炖____________
7. beef shunk ____________ 8. 切碎的末____________
9. tenderloin ____________ 10. 鸡翅____________

Exercise 2 Tick off the different words.(找不同)

1. A. slice B. simmer C. shred D. cube
2. A. steam B. marinate C. mince D. stir-fry
3. A. steam B. boil C. deep-fry D. dice
4. A. sugar B. cooking wine C. starch D. ginger
5. A. chunk B. pork round C. spare ribs D. chicken thigh

Exercise 3 Choose the best answers.(单项选择)

1. What ____________ we going to learn?
A. shall B. are C. is D. am
2. There are different ways of cutting, ____________ there?
A. isn't B. aren't C. don't D. doesn't
3. Cut the tenderloin ____________ filets and mix them ____________ salt and egg white.
A. into;with B. with;into C. from;in D. from;to
4. —Look! I'll show you ____________ to make it. —That's wonderful!
A. what B. where C. how D. when
5. I know kung pao chicken is ____________ than the other dishes.

A. more popular B. the most popular

C. the more popular D. most popular

Exercise 4 Fill in the blanks.（参考例子，补全对话）

eg
What shall I do with the beef?
What shall I do with the chicken?

1. Why not cook the diced chicken for a long time?

________ ________ ask Beijing roast duck for lunch.

2. Could you tell me how to cut beef?

________ you ________ me how to make kung pao chicken.

3. Both the chili and watercress make the dish very spicy.

________ she ________ I like sweet and sour pork with pineapple.

4. I'll show you how to scramble the eggs.

I'll ________ you ________ to stir-fry the tenderloin.

Exercise 5 Write.（写出下列菜肴的英文表达）

Dongpo elbow	paprika chicken	sautéed beef with scallion
steamed chicken	roast leg of lamb	lamb with cumin

1. 白斩鸡____________ 2. 辣子鸡____________ 3. 东坡肘子____________

4. 烤羊腿____________ 5. 孜然羊肉____________ 6. 葱爆牛肉____________

Exercise 6 Translate.（中英文句子翻译）

1. 我该怎样烹制宫保鸡丁？

____

2. 用长柄刀将骨头从肉上切下来。

____

3. Add dried red chili，chopped scallion，sliced ginger and white pepper to stir flavor.

____

4. We can cut meat into sliced，shredded，blocked，cubed and minced shapes.

____

5. Put fried chicken breast and seasoning sauce for stir-fry.

____

Unit 6

参考答案

Note

Unit 7
Seafood

Learning goal（学习目标）

You will be able to：

1. know the words and expressions of seafood；
2. describe the ways of trimming fish and lobster；
3. talk about the steps of making seafood dishes.

扫码看课件

Lead in(导入)

Activity 1 Look and match. 看图并连线。

Activity 2 Decide true (T) or false (F). 判断正误。

1. ________ Eel is nutritious(有营养的), which fits for children and the old people.
2. ________ Carps' jumping over the dragon gate is a legend(传说) handed down in China for many years.
3. ________ Shrimps and clams belong to shellfish.
4. ________ Tunas can be made into cans.
5. ________ A crab has eight legs, one head and two wings.

Student learning(学生学习)

Task 1 Words & expressions 词汇

Task description. 任务描述。

水产品包括鱼、虾、蟹、贝类，营养丰富，味道鲜美，是人类所需动物性蛋白质的重要来源。水产品大致分成海水鱼类、淡水鱼类和贝壳类等。任务一将聚焦于海水鱼、淡水鱼和虾贝蟹类的单词和词组的学习。

Note

Part 1 Sea fish. 海水鱼。

外教有声

Activity 1 Read. 读单词和词组。

pomfret [ˈpʌmfrɪt] 平鱼
yellow croaker [ˈkrəʊkə(r)] 黄花鱼
grouper [ˈgruːpə(r)] 石斑鱼
saury [ˈsɔːrɪ] 秋刀鱼
Spanish mackerel [ˈmækərəl] 鲅鱼
codfish [ˈkɒdfɪʃ] 鳕鱼
sea bass [beɪs] 海鲈鱼
ribbonfish [ˈrɪbənfɪʃ] 带鱼
turbot [ˈtɜːbət] 多宝鱼

Activity 2 Look and write. 看图写中文。

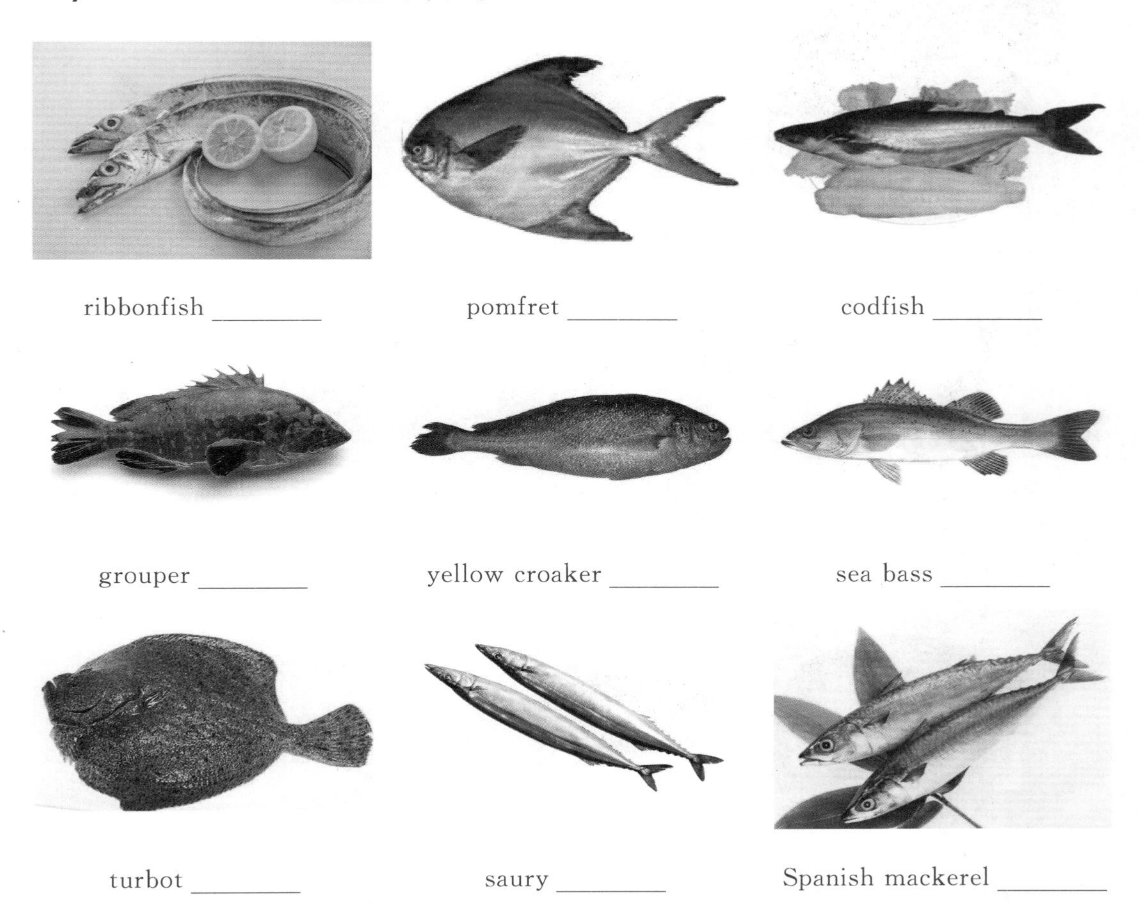

ribbonfish ________ pomfret ________ codfish ________

grouper ________ yellow croaker ________ sea bass ________

turbot ________ saury ________ Spanish mackerel ________

Note

Part 2 Fresh-water fish. 淡水鱼。

外教有声

Activity 1 Read. 读单词和词组。

crucian ['kruːʃən] 鲫鱼

grass [grɑːs] carp 草鱼

silverfish['sɪlvəfɪʃ] 银鱼

mackerel ['mækərəl] 鲭鱼

mandarin ['mændərɪn] fish 鳜鱼

catfish ['kætfɪʃ] 鲇鱼

silver carp 鲢鱼

trout [traʊt] 鳟鱼

Activity 2 Look and write. 看图写英文。

鲫鱼________ 鳜鱼________ 鳟鱼________

草鱼________ 银鱼________ 鲭鱼________

鲇鱼________ 鲢鱼________

Note

Part 3 Shrimps & shellfish. 虾贝蟹类。

外教有声

Activity 1 Read. 读单词。

lobster ['lɒbstə(r)] 龙虾
prawn [prɔːn] 对虾，明虾
scallop ['skɒləp] 扇贝
oyster ['ɔɪstə(r)] 牡蛎，生蚝
mussel ['mʌsəl] 贻贝，青口
abalone [ˌæbə'ləʊnɪ] 鲍鱼
sea snail [sneɪl] 海螺
king crab 帝王蟹

Activity 2 Write and translate. 单词互译。

扇贝 ________

龙虾 ________

对虾 ________

鲍鱼 ________

oyster ________

mussel ________

king crab ______

sea snail ______

Activity 3 Complete. 补全单词。

1. s __ ury
2. sc __ llop
3. __ b __ lone
4. s __ __ sn __ il
5. m __ nd __ rin
6. l __ bst __ __
7. c __ dfish
8. p __ mfr __ t
9. tr __ __ t
10. __ yst __ __

Note

Activity 4 Making dialogues. 对话练习。

Chef: What seafood would you like to eat?

Jack: I'd like to eat ________ .

Task 2 Trimming & keeping 水产品的初加工和保鲜

Task description. 任务描述。

水产品种类繁多，加工方法不尽相同。随着人们生活水平的提高，越来越多的人喜欢水产品，因此水产品的保鲜十分重要。任务二将聚焦于鱼类和虾蟹类产品的初加工以及它们的保鲜方法的英文表达。

Part 1 Trimming fish. 鱼的初加工。

外教有声

Activity 1 Read. 读单词和词组。

fin [fɪn] 鱼翅

cavity ['kævətɪ] 腔，洞

scrape [skreɪp] off 刮掉，擦去

fish scissors ['sɪzəz] 鱼剪

gut [gʌt] 肠子，内脏

blood [blʌd] 血，血液

fish scaler ['skeɪlə(r)] 刮鳞器

shallow incision [ɪn'sɪʒən] 浅的切口

Note

外教有声

Activity 2 Listen and complete. 听录音，补全对话。

A. fish scaler and fish scissors	B. shallow incision
C. scrape the scales off	D. remove the guts

Chef: Today I'll teach you how to trim the fish.

Jack: Wonderful! What do we call these fish tools in English, chef?

Chef: They are ________.

Jack: How shall we do with them?

Chef: You can ________ with a fish scaler and remove the fins with a pair of fish scissors.

Jack: I've done it. What's the next?

Chef: Make a ________ in the underside from just before the fin to the head and ________ using your hands.

Jack: I've finished. Shall I wash it now?

Chef: Yes, rinse the cavity with cold running water to clean the remaining blood and guts.

Jack: I've got it. Thank you!

Activity 3 Write. 写出加工鱼所用工具的英文表达。

________ ________ ________

Activity 4 Look and choose. 看图并选择。

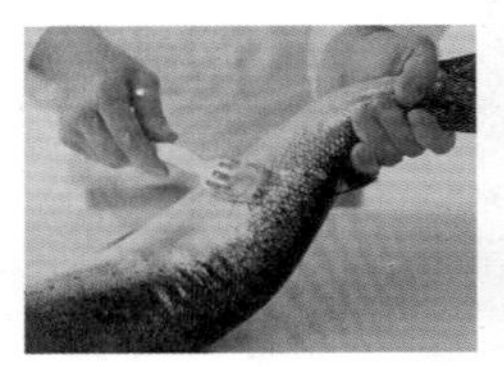

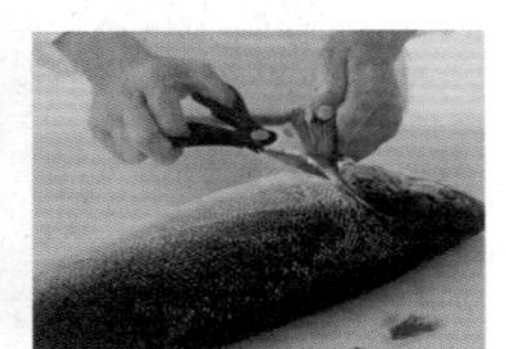

________ ________

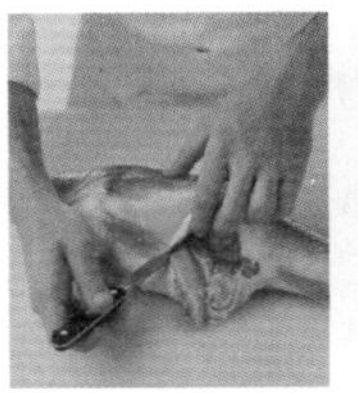

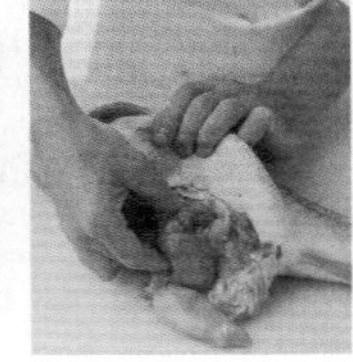

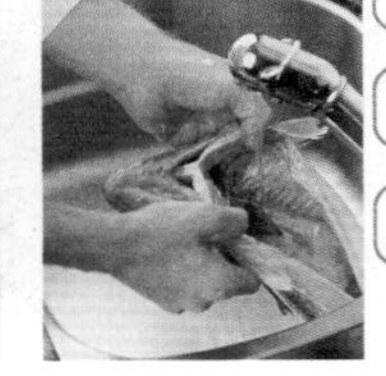

________ ________ ________

A. Rinse the cavity with cold running water.

B. Make a shallow incision in the underside.

C. Scrape the scales off with a fish scaler.

D. Remove the guts using the hands.

E. Remove the fins with a pair of fish scissors.

Note

Part 2 Trimming lobster. 虾的初加工。

外教有声

Activity 1 Read. 读单词和词组。

twist [twɪst] 拧,扭曲	thumb [θʌm] 拇指
remove [rɪˈmuːv] 移动,搬走	hammer [ˈhæmə(r)] 锤子
take hold of the tail 握住尾巴	cut down the center 切开中部
pull... apart 分开	lobster cracker 龙虾钳

外教有声

Activity 2 Listen and write. 听录音,写出听到的内容。

Jack: What shall we do with the lobster before cooking, chef?

Chef: We should ____________ and twist it.

Jack: OK. It's so easy. And then?

Chef: ____________ with the sharp kitchen scissors.

Jack: What's the next?

Chef: Pull the shell apart ____________.

Jack: Shall we use the ____________ next?

Chef: Yes, we can ____________ from them.

Jack: Could I use a small hammer instead of lobster cracker?

Chef: No problem.

Activity 3 Look and describe. 参考 **Activity 2** 用英文描述下列图片。

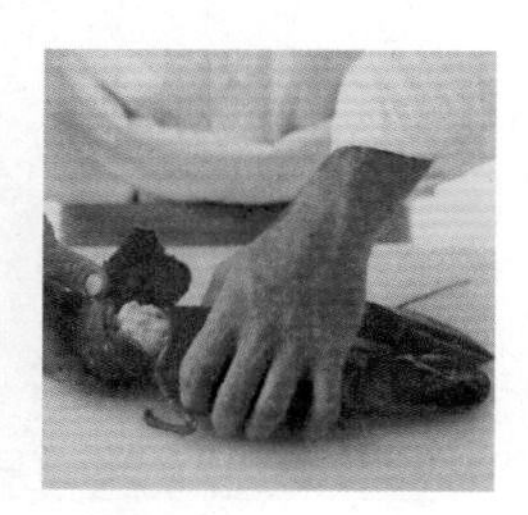

____

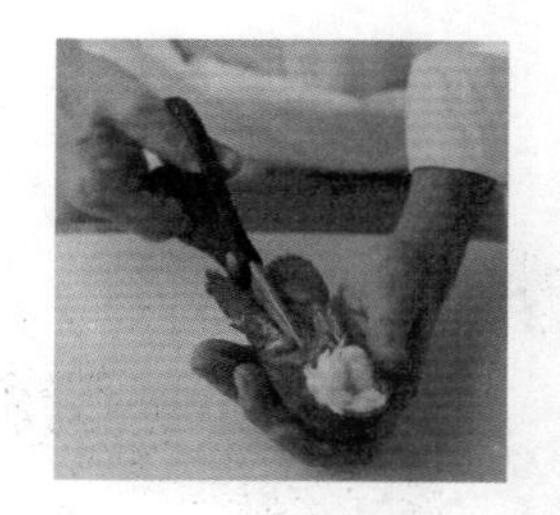

____

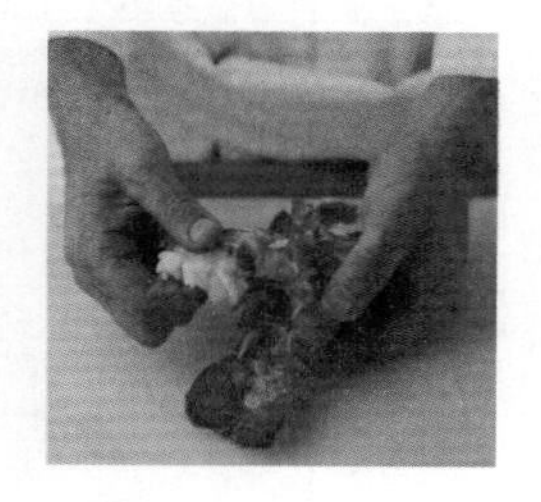

____

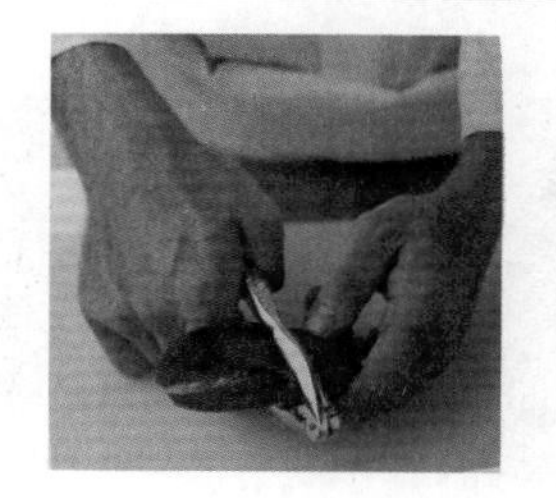

____

Note

Keeping freshness. 水产品的保鲜。

Activity 1 Look and match. 将鱼类保存的方法与图片连接。

A. Cut the fish into pieces and keep them with tissue or freshness protection package.

B. Scale the fish and remove the guts.

C. Keep them in the refrigerator.

Activity 2 Translate. 将虾类保存的方法翻译成中文。

1. Wash the shrimps in running water.

____

2. Put them in a metal basin and keep in freezer.

____

3. Take out the basin and leave it outside for a while.

____

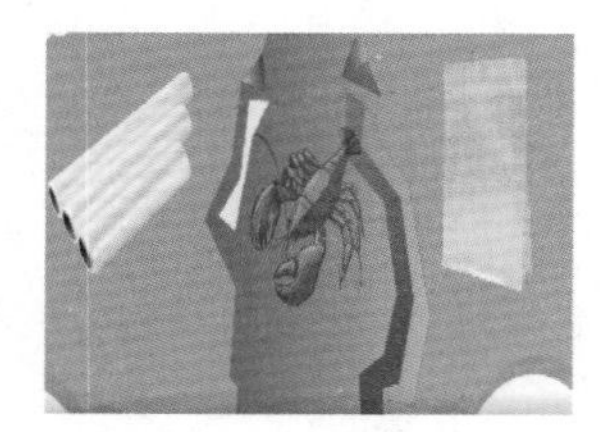

4. Pour out the frozen ice and pack with freshness protection package.

____

5. Put them in the refrigerator.

____

Task 3 Making dishes 菜肴制作

Task description. 任务描述。

水产品含有丰富的蛋白质、钙、牛磺酸、磷、维生素 B_1 等多种人体所需的营养成分，脂肪含量极低。任务三将聚焦于剁椒鱼制作过程中所需要的原料和加工过程的英文表达。

外教有声

Activity 1 Read the recipe. 读菜单。

Steamed fish with chopped red chili (剁椒鱼)

Ingredients(原料)

a grass carp, around 500 g (it has been cleaned of scales and inwards)

chopped red chili green onion sections

chopped garlic ginger slices salt

chicken essence (chicken stock) cooking wine

five spices powder

Methods (制作方法)

Note

1. Add in the green onion sections, ginger slices, cooking wine, chicken essence and salt.

Marinate(腌制) the grass carp for 15-20 minutes.

2. Add in the chopped garlic and chopped red chili. Stir-fry for about 20 seconds until fragrant, and then turn off the heat.

3. Put the stir-fried chopped red chili on the fish.

4. Put the plate on the basket and cover it. Turn on the heat to steam it for 8-10 minutes.

5. Turn off the heat and steam it for another 5 minutes without heat. Pick it out and garnish it.

Characteristics(特点)

It has a strong taste because of the spicy red chili, but rather tender and fresh.

Activity 2 Decide true (T) or false (F). 判断正误。

1. ________ Steamed fish with chopped red chili is Guangdong dish.
2. ________ Steamed fish and poached fish have different tastes.
3. ________ Cooking wine is not necessary in this dish.
4. ________ We usually serve dish in the mixing bowl.
5. ________ The fish will be fried on both sides.

Activity 3 Translate. 中英文互译。

Ingredients (原料)

草鱼____________

辣椒块____________

葱段____________

蒜末____________

ginger slices ________

chicken essence ________

salt ____________

cooking wine ________

five spices powder ________

Methods（制作方法）

1. Marinate the fish with the green onion sections, ginger slices, cooking wine, chicken essence and salt for 15-20 minutes.

____

2. Add in the chopped garlic and chopped red chili. Stir-fry for about 20 seconds until fragrant.

____

3. Put the stir-fried chopped red chili on the fish.

____

4. Put the plate on the basket and cover it. Steam it for 8-10 minutes.

____

5. Turn off the heat and steam it for another 5 minutes without heat. Pick it out and garnish it .

____

Culture knowledge（餐饮文化）

吃海鲜的饮食禁忌

吃海鲜的时候需要注意很多问题,下面我们一起来看看海鲜怎么吃才健康吧。

Note

一、海鲜不能和什么一起吃

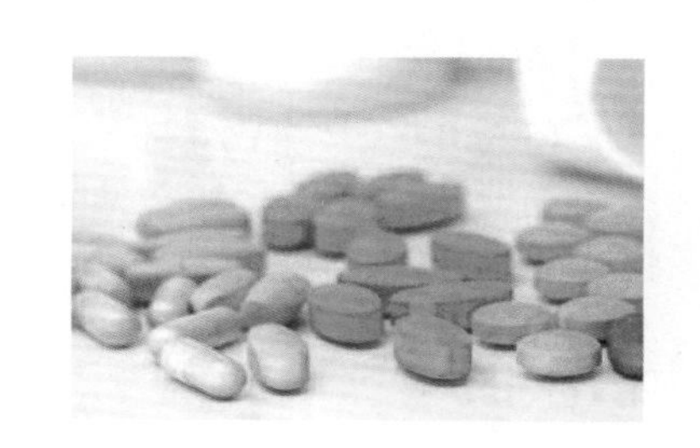

❶ 海鲜和维生素 C 同食会中毒

多种海鲜，如虾、蟹、蛤蜊、牡蛎等，体内均含有化学元素砷。一般情况下含量很低，但日益严重的环境污染可能使这些动物体内砷的含量达到较高水平，导致人体中毒。

❷ 海鲜不要和水果混着吃

海鲜中含有较丰富的蛋白质、钙、磷等营养素，如果与鞣酸含量较高的食品一起吃，会使海鲜中的钙类与鞣酸结合，导致呕吐、头晕、恶心、腹痛等症状。因此，茶叶、柿子、葡萄、山楂等含鞣酸较多的水果，不能与海鲜同食，至少要隔开 2 小时。

❸ 海鲜啤酒同吃惹痛风

虾、蟹等海鲜在体内代谢后会形成尿酸，而尿酸过多会引起痛风、肾结石等病症。如果在大量食用海鲜的同时饮用啤酒，就会加速体内尿酸的形成。所以，在大量食用海鲜的时候，千万别喝啤酒，否则会对身体产生不利影响。

二、海鲜怎么吃健康

❶ 海鲜需要煮熟灭菌

海鲜中的病菌主要是副溶血性弧菌等，耐热性比较强，80 ℃以上才能杀灭。因此，在吃“醉蟹”“生海胆”“酱油腌海鲜”之类不加热烹调的海鲜时一定要慎重，吃生鱼片的时候也要保证鱼的新鲜和卫生。

❷ 贝类海鲜不宜存放太久

不新鲜的贝类会产生较多的胺类和自由基，对人体健康造成威胁。选购活贝之后也不能存放太久，要尽快烹调。

❸ 打包来的海鲜要冷藏

如果海鲜已经高温彻底烹熟，只需马上放入冷藏室，下一餐加热后即可食用；如果海鲜并未经过充分加热，应当放进冷冻室，下一餐之前化冻，彻底加热烹熟。由于海鲜类食品的蛋白质质地细腻、分解很快，拿回家之后应当在一天之内吃完，不要长时间存放。

Unit exercise（单元练习）

Exercise 1 Translate.（单词互译）

1. catfish ____________　　2. 龙虾____________

3. codfish ____________ 4. 贻贝 ____________

5. ribbonfish ____________ 6. 鲍鱼 ____________

7. silver carp ____________ 8. 海螺 ____________

9. mackerel ____________ 10. 草鱼 ____________

Exercise 2 Tick off the different words. (找不同)

1. A. grouper	B. saury	C. turbot	D. silverfish
2. A. shrimp	B. mussel	C. trout	D. oyster
3. A. take out	B. cut open	C. scale	D. scallop
4. A. delicious	B. terrible	C. gut	D. wonderful
5. A. vinegar	B. sugar	C. soy sauce	D. crab

Exercise 3 Match. (连线)

1. What can I use instead of lobster cracker?	A. Scale it with the fish scaler.
2. What ingredients do we need?	B. Ginger slices, green onion sections, sugar, vinegar and soy sauce.
3. What is the best way to clean it?	C. You can use hammer instead of it.
4. How long shall I stew the fish?	D. Wash it in salted water.
5. How shall I do with the fish?	E. Cover and stew it for 8 minutes.

Exercise 4 Put words in right order. (句子排序)

1. on, the, fish, fry, for, 1 minute, both sides, about

____

2. on, fish, the, stir-fried, chopped, put, red chili

____

3. hands, using, remove, the, guts, your

____

4. fish, trim, the, I, teach, you, how, to, will

____

5. what, you, eat, of, kind, would, shellfish, like, to

____

Note

Exercise 5 Write.（写出下列菜肴的英文表达）

West Lake fish in vinegar gravy	steamed Wuhan fish
braised ribbonfish in brown sauce	sautéed shrimps with broccoli
sautéed crab in hot spicy sauce	sweet and sour mandarin fish

1. 松鼠鳜鱼_______________　2. 香辣蟹_______________　3. 红烧带鱼_______________

4. 西湖醋鱼_______________　5. 清蒸武昌鱼_______________　6. 翡翠虾仁_______________

Exercise 6 Translate.（将贝类保存方法翻译成英文）

1. 用清水清洗。

____

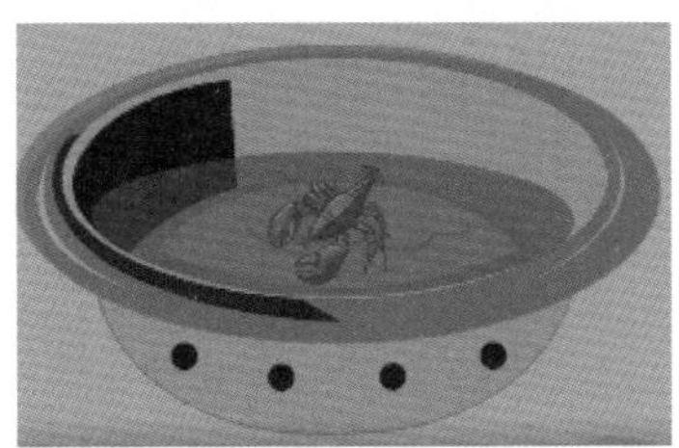

2. 浸泡在盐水中。

____

3. 放在冰箱中冷冻。

____

Exercise 7 Translate.（中英文句子翻译）

1. 不同的工具有不同的用途。

____

2. 把鱼切成小块用保鲜袋包裹后进行保存。

____

3. 麦克和他的师傅正在做清蒸鲈鱼。

____

4. Stir-fry the ingredients for about 20 seconds until fragrant.

____

5. Remove the grass carp from the wok and serve on a plate.

____

Unit 7

参考答案

Note

Unit 8

Chinese Pastries

Learning goal（学习目标）

扫码看课件

You will be able to:

1. get familiar with the words and expressions of Chinese pastries;
2. know the tools and materials of making Chinese pastries;
3. describe the steps of making Chinese pastries.

Lead in(导入)

Activity 1 Look and match. 看图并连线。

- steamed bun (mantou) 馒头
- porridge ['pɒrɪdʒ] 粥
- rice [raɪs] 米饭
- noolde ['nuːdl] 面条
- pancake ['pænkeɪk] 煎饼
- deep-fried dough sticks (youtiao) 油条

Activity 2 Decide true (T) or false(F). 判断正误。

1. ________ Noodles are the main food in the southern Chinese diet.
2. ________ Porridge and youtiao are common breakfasts in Beijing.
3. ________ Noodles are made from wheat flour.
4. ________ Rice is a major staple food for people in southern China.
5. ________ Youtiao is a kind of fried staple food.

Student learning(学生学习)

Task 1 Words & expressions 词汇

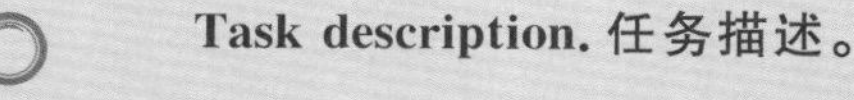

Task description. 任务描述。

中式面点是我国烹饪的主要组成部分,历史悠久,制作精致,品类丰富,风味多样,基本分类方法也很多,按照形状可分成包类、饺类、糕类、卷类、面条类、饼类、饭类、粥类和汤等。任务一将聚焦于各类面点的单词和词组的学习。

Note

Part 1 Steamed stuffed buns & dumplings. 包类和饺类。

外教有声

Activity 1 Read. 读单词和词组。

cha siu bao 叉烧包

shaomai 烧卖

wonton [wɒnˈtɒn] 馄饨

small steamer bun 小笼包

fried dumpling 锅贴

steamed creamy custard bun [ˈkriːmɪ ˈkʌstəd bʌn] 奶黄包

steamed shrimp dumpling [ˈdʌmplɪŋ] 虾饺

smashed bean [smæʃt biːn] bun 豆沙包

pan fried pork bun 生煎包

Activity 2 Look and write. 看图写中文。

shaomai ________ steamed creamy custard bun ________

cha siu bao ________

steamed shrimp dumpling ________

wonton ________

smashed bean bun ________

small steamer bun ________

pan fried pork bun ________

fried dumpling ________

Note

Part 2 Cakes & rolls. 糕类和卷类。

外教有声

Activity 1 Read. 读词组。

cotton ['kɒtən] cake 棉花糕
horseshoe ['hɔːsʃuː] cake 马蹄糕
green bean cake 绿豆糕
rice cake 年糕
sweet soup [suːp] balls 汤圆
sweet dumplings 元宵
fried rice balls with sesame ['sesəmɪ] 麻团
silver thread roll [rəʊl] 银丝卷
egg roll 蛋卷

Activity 2 Look and write. 看图写英文。

元宵________　蛋卷________　绿豆糕________

棉花糕________　年糕________　麻团________

马蹄糕________　汤圆________　银丝卷________

Note

Part 3 Noodles & pancakes. 面条类和饼类。

外教有声

 **Activity 1** Read. 读词组。

beef noodles 牛肉面
plain [pleɪn] noodles 阳春面
sliced noodles 刀削面
corn pancake 玉米饼
beef pancake 牛肉饼
noodles with soy bean paste 炸酱面
hot dry noodles 热干面
egg pancake 鸡蛋饼
fried baked scallion pancake 葱油饼

Activity 2 Write and translate. 单词互译。

牛肉面 ________________

葱油饼 ________________

炸酱面 ________________

热干面 ________________

corn pancake ________________

plain noodles ________________

sliced noodles ________________

beef pancake ________________

Note

Part 4 Rice, porridge & soup. 饭类、粥类和汤。

外教有声

Activity 1 Read. 读词组。

Yangzhou fried rice 扬州炒饭
braised [breɪzd] chicken with rice 黄焖鸡米饭
millet congee [ˈmɪlɪt ˈkɒndʒiː] 小米粥
vegetable soup 蔬菜羹
preserved egg and pork porridge 皮蛋瘦肉粥
fried rice in soy sauce 酱油炒饭
pumpkin porridge 南瓜粥
West Lake beef soup 西湖牛肉羹

Activity 2 Look and write. 看图写英文。

Activity 3 Complete. 补全单词。

1. n __ __ dle 2. p __ nca __ e 3. m __ l __ et 4. p __ __ ri __ ge 5. __ un
6. pl __ __ n 7. du __ pl __ ng 8. won __ __ n 9. st __ __ me 10. con __ e __

Note

Activity 4 Making dialogues. 对话练习。

A: What Chinese pastry would you like to eat?

B: I'd like to eat ________ .

Task 2 Making pastries 面食制作

Task description. 任务描述。

中国的面食种类繁多，制作方法也百花齐放。随着人们生活水平的提高，越来越多的人喜欢了解面食的制作及其方法。面食从熟制方法上可以分为蒸、煮、烙、煎、烤、炸、焖等类型。制作的方法不同，营养素损失也有所不同。任务二将聚焦于面食制作的原料、工具、方法等的英文表达。

Part 1 Materials & tools. 加工原料和工具。

Activity 1 Read. 读单词和词组。

外教有声

high gluten flour ['gluːtən 'flaʊə(r)] 高筋粉

low gluten flour 低筋粉

sweet rice flour 糯米粉

baking powder ['paʊdə(r)] 泡打粉

yeast [jiːst] powder 酵母粉

moulage [muː'lɑːʒ] 印模

rolling pin [pɪn] 擀面杖

scraper ['skreɪpə(r)] 刮面板

Note

Activity 2 Look and write. 看图写词组。

____

____

____

____

____

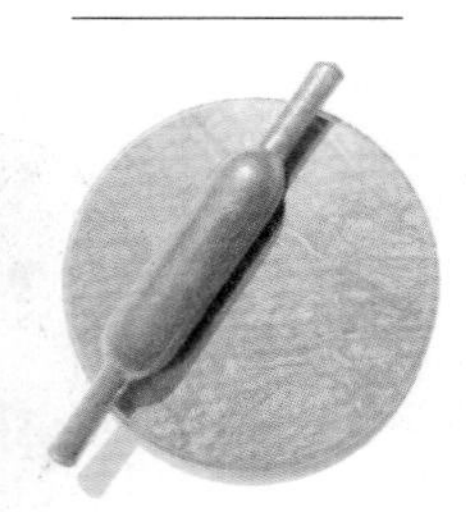

____

____

____

外教有声

Activity 3 Listen and complete. 听录音,补全对话。

A. Place high gluten flour in a bowl

B. add a tea spoon of yeast powder

C. divide the dough with scraper

D. roll into a round by using rolling pin

Chef: Let's get ready for making dumplings.

Jack: Wonderful! But what shall we do on the first step, chef?

Chef: Flour first.

Jack: There are two kinds of flour. Which one is better for dumplings?

Chef: High gluten flour. ______________. Add some water slowly. Mix with fingers to form a soft dough. And then ______________ in it.

Jack: I've got it. How many minutes does it need to stand for?

Chef: 15 minutes. Next, ______________.

Jack: Last one I know, flatten each piece and ______________.

Chef: OK, the preparation is done!

Note

Activity 4 Look and choose. 看图并选择。

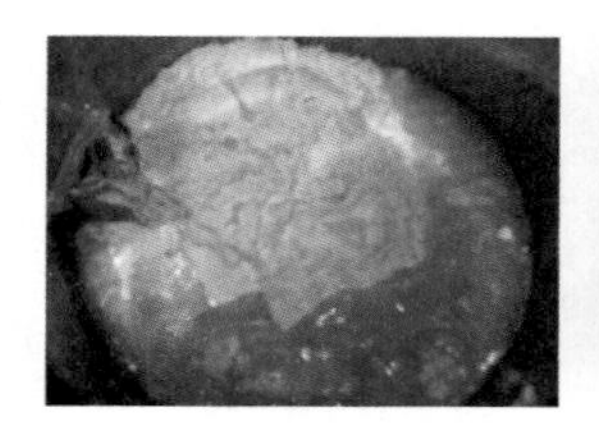

____

____

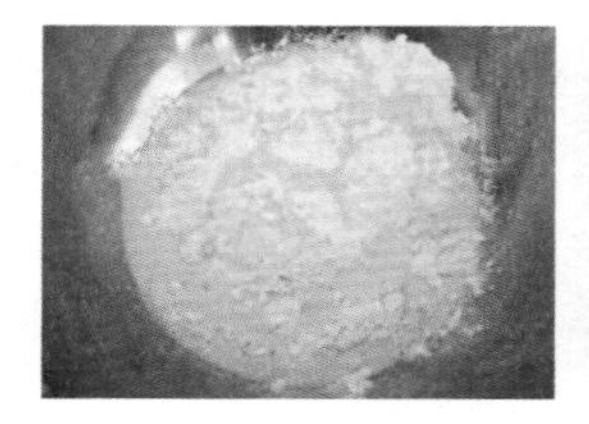

____

____

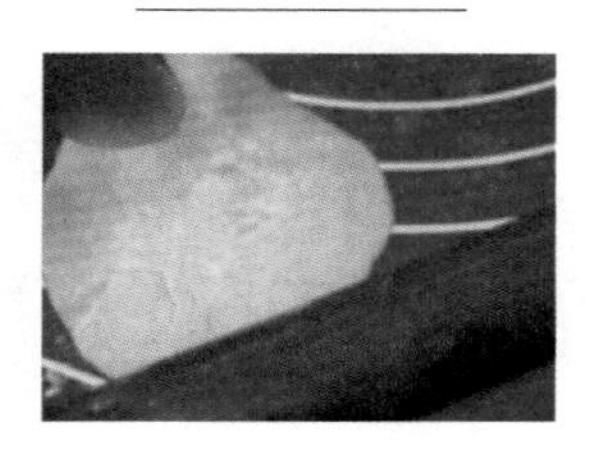

____

A. Place high gluten flour in a bowl.

B. Add a tea spoon of yeast powder.

C. Divide the dough with scraper.

D. Roll into a round by using rolling pin.

E. Stand for 15 minutes.

Part 2 Making dumplings. 饺子的制作。

Activity 1 Read. 读词组。

外教有声

let the dough rest for 10 minutes 醒面 10 分钟

settle together 融合

mix the flour, salt and water in a large bowl 碗里放入面粉、盐和水

use the rolling pin 用擀面杖

Activity 2 Listen and write. 听录音，写出听到的内容。

外教有声

Jack: Could you tell me how to make steamed shrimp dumplings?

Chef: No problem. First, we need to ________.

Jack: What is the next?

Chef: We will ________ because it can help ingredients ________.

Jack: A cutting board will be working here.

Chef: Yes. Cut the dough into about 20 or more equal pieces for the dumplings.

Jack：I see. You told me ________ to smooth out each disc until it likes a small pancake.

Chef：Wonderful! Then put some filling in the middle of the round wrapper and pack it.

Chef：Steam the dumplings for 10 to 15 minutes until they can be served.

Activity 3 Look and match. 将炒面的制作方法与图片连接。

A. Add the boiled noodles into mixture of the vegetables.

B. Boil the noodles.

C. Chop all the vegetables into long strips.

D. Fry the vegetables together for 4 minutes.

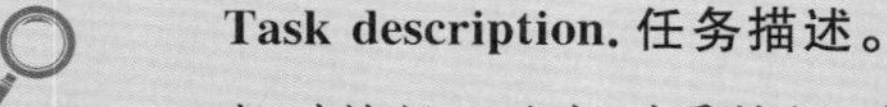

Task 3 Making fried rice 炒饭的制作

Task description. 任务描述。

扬州炒饭又名扬州蛋炒饭，制作精良，加工讲究，注重配色，是江苏经典小吃。任务三将聚焦于扬州炒饭的原料和制作过程的英文描述的学习。

Note

外教有声

 **Activity 1** Read the recipe. 读菜单。

Yangzhou fried rice(扬州炒饭)

Ingredients(原料)

an egg　50 g diced sausage　20 g boiled shelled shrimps　50 g diced carrot

20 g boiled green beans　a bowl of cooked rice　chopped scallion

salt and white pepper

Methods (制作过程)

1. Heat oil in the wok. Drop in the egg and fry it, and then set aside.

2. Add in the chopped scallion, diced carrot, boiled green beans, diced sausage and boiled shelled shrimps. Stir-fry them.

3. Add in cooked rice, and stir all together. Then pour in the egg and mix them.

4. Season with salt and white pepper to taste.

Characteristics (特点)

It is known as the careful material selection and precise cooking, and featuring with characteristics of Huaiyang cuisine.

Activity 2 Decide true (T) or false (F). 判断正误。

1. ________ Yangzhou fried rice is a Beijing dish.
2. ________ Steamed rice and fried rice are different cooking ways.
3. ________ Sugar is not necessary in this dish.
4. ________ Refrigerated rice will make your fried rice more perfect.

Activity 3 Translate. 中英文互译。

Ingredients(原料)

火腿丁________

虾仁________

鸡蛋________

青豆________

salt ________

white pepper ________

chopped scallion ________

diced carrot ________

Note

Methods(制作过程)

1. Heat oil in the wok. Drop in the eggs and fry them.

2. Add in the chopped scallion, diced carrot, boiled green beans, and then stir-fry them.

3. Add in cooked rice, and stir all together. Then pour in the eggs and mix them.

4. Season with salt and white pepper to taste.

Culture knowledge(餐饮文化)

七大中式传统糕点派系

❶ 京派

以北京地区为代表。历史悠久，品类繁多，滋味各异，具有重油、轻糖，酥松绵软，口味纯甜、纯咸等特点。代表品种有京八件和红、白月饼等。

❷ 苏派

以苏州地区为代表。馅料多用果仁、猪板油丁，用桂花、玫瑰调香，口味重甜。代表品种有苏式月饼和猪油年糕等。苏式糕点在中国汉族糕点发展史上占有重要地位。

❸ 广派

以广州地区为代表。用料精博，品种繁多，款式新颖，口味清新多样，制作精细，咸甜兼备，并糅合了西式糕点的技巧和特色。代表品种有绿茵白兔饺、煎萝卜糕、马蹄糕、皮蛋酥等。

❹ 潮派

以潮州地区为代表。馅料以豆沙、糖冬瓜、糖肥膘为主，葱香风味突出。代表品种有老婆饼和春饼。

⑤ 宁派

以宁波地区为代表。宁式糕点选料讲究，并形成以酥为主，酥、软、脆分明的特点。极有特色的苔生片、苔菜千层酥、苔菜月饼，以及松脆香甜的多孔"三北"藕丝糖是其代表。

⑥ 川派

以成渝地区为代表。馅料多用花生、芝麻、核桃、蜜饯、猪板油丁，糯米制品较多，软糯油润，香甜酥脆。代表品种有桃片和米花糖等。

⑦ 滇派

以云南地区为代表。选取富有当地特色的原始材料精心烹饪而成，鲜美可口，回味无穷。代表品种有云腿月饼和鲜花饼等。

Unit exercise（单元练习）

Exercise 1 Translate.（单词互译）

1. hot dry noodles ____________
2. plain noodles ____________
3. spring rolls ____________
4. 饺子____________
5. beef noodles ____________
6. 炒面____________
7. 马蹄糕____________
8. 馄饨____________
9. yeast powder ____________
10. 烧卖____________

Exercise 2 Write the right order.（写出做虾饺的正确顺序）

A. Prepare the dumplings filling.

B. Put some filling in the middle of the round wrapper and pack it.

C. Make a dumpling dough.

D. Roll each dumpling ball into around wrapper.

E. Steam shrimp dumplings for 10 to 15 minutes until they are done.

1. ________　2. ________　3. ________　4. ________　5. ________

Exercise 3 Choose the best answer.（单项选择）

1. Place the dough on a lightly floured board and roll it ________.

A. on　　B. out　　C. up　　D. of

2. At last, the chef will sprinkle the scallion ________ top.

A. up　　B. on　　C. in　　D. above

3. Please ________ each ball into a round wrapper.

A. add　　B. form　　C. bake　　D. put

4. Divide the mixture into walnut size balls after ________ well.

A. is kneading B. kneading C. knead D. to knead

5. Onions and spices may be ________ to make your rice savory.

A. add B. adding C. added D. to add

6. What kind of noodles shall I ________?

A. make B. cut C. study D. look

7. Can you tell me how to ________ scrambled eggs?

A. making B. made C. make D. be made

8. First, let's heat the oil ________ the wok.

A. on B. in C. of D. to

9. Pour ________ the eggs, mix them. Season ________ salt to taste.

A. in; of B. on; to C. in; with D. to; with

10. Cover the pan and ________ the heat to very low.

A. place B. reduce C. break D. mix

Exercise 4 Write.(写出下列面点的英文表达)

sliced noodles　　cold noodles with sesame sauce　　steamed twisted roll
fried rice with seafood　　fried baked scallion pancake　　wotou with black rice

1. 凉面________ 2. 葱油饼________ 3. 刀削面________

4. 海皇炒饭________ 5. 花卷________ 6. 黑米小窝头________

Exercise 5 Translate.(中英文句子翻译)

1. 蛋炒饭做好了。

____

2. 把拌好的馅用饺子皮包好。

____

3. 我们用番茄和鸡蛋做什么?

____

4. Add the green peas, diced carrot and salt, and stir well.

____

5. Please break three eggs in the bowl.

____

Unit 8
参考答案

Note